AF894697

53 Graphic Organizers for Science

Sunflower education

Exceptional Books for Teachers and Parents

Illustrators Fian Arroyo
Steven Durke

ISBN-13: 978-1-937166-26-7

Permission to Reproduce
Permission is granted for the user to reproduce the designated blackline masters. Reproducible pages are indicated by the symbol on the left side of the bottom margin. The reproduction of any part of this program for commercial use or for an entire school system, hospital system, or institutional system is strictly prohibited.

Contents

Contents

Introduction

Graphic organization of information comes naturally to people. It even predates writing. Ancient peoples used a variety of ingenious graphic methods to organize, record, and transmit information for centuries before some of these methods developed into writing. This fact alone seems to indicate that the human mind is well-suited to organizing and interpreting information presented graphically.

Modern psychological and educational research has proved the effectiveness of different visual devices—especially graphic organizers—as powerful communication and learning tools. Indeed, dozens of studies have established that graphic organizers provide students and teachers with numerous benefits.

Benefits for Students

Recent scientific studies have only confirmed what teachers working in the trenches have known for some time: graphic organizers help students. You can ask any teacher with experience using graphic organizers, and he or she will likely identify at least one, and probably several, benefits of graphic organizers for students:

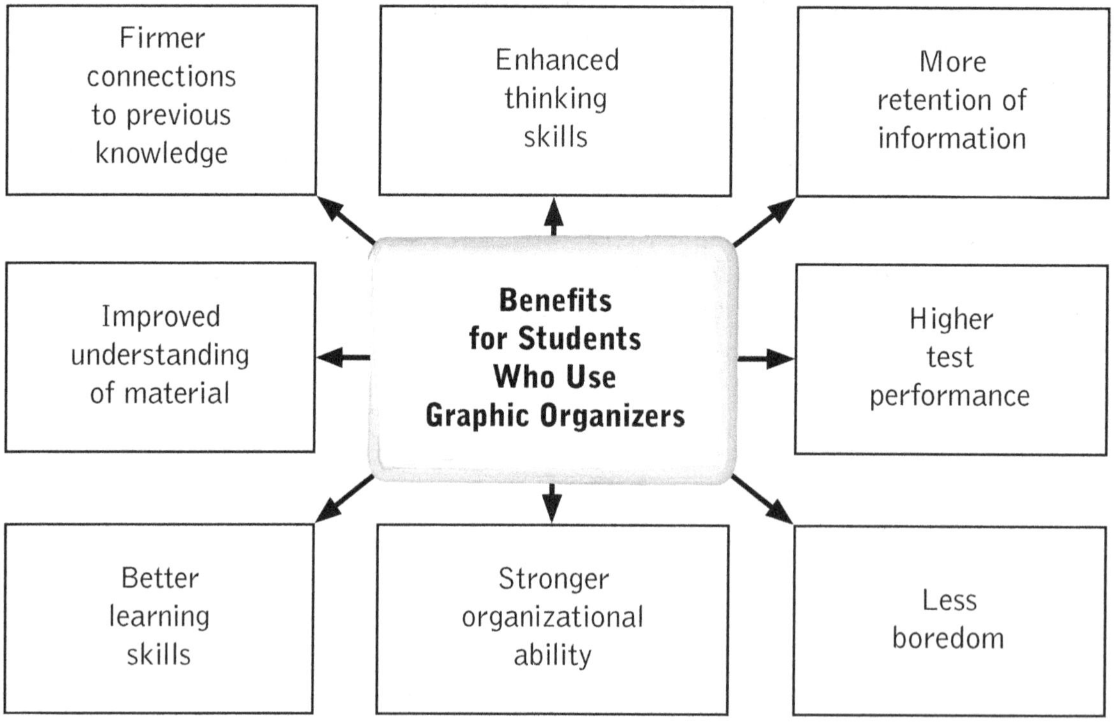

Benefits for Teachers

Teachers who use graphic organizers benefit as much as the students. The overriding advantage is that you will elicit better performance from your students. In addition, the following benefits have also been reported:

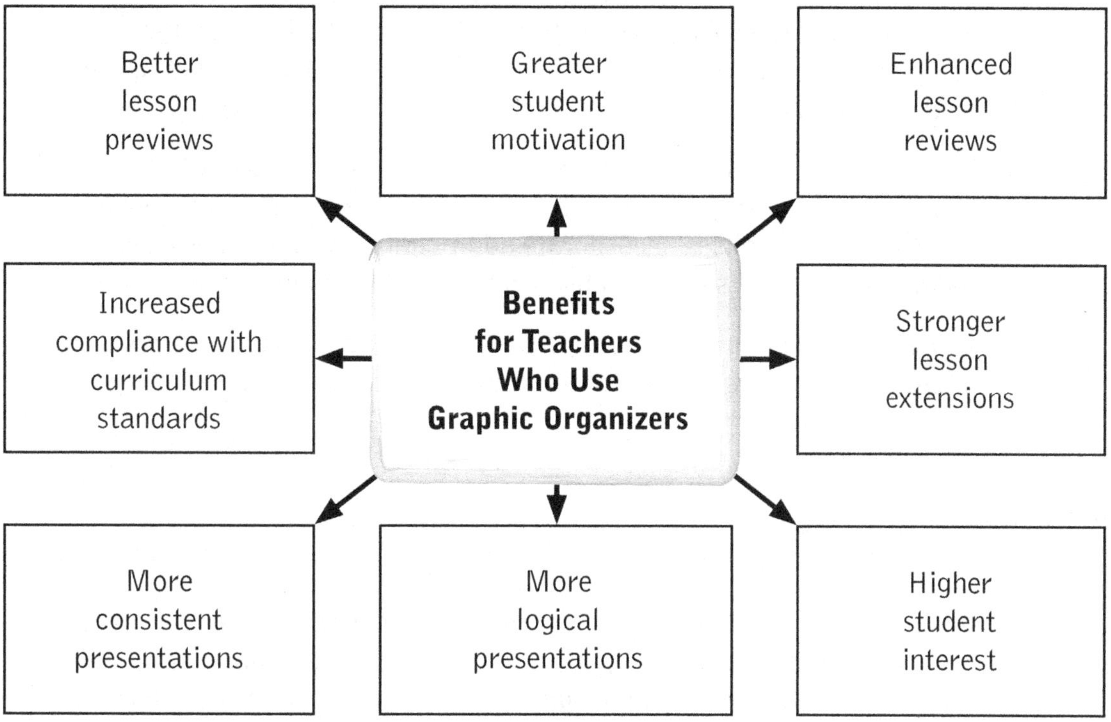

Using *Graphic Organizers: Science*

The graphic organizers in this book were conceived, designed, and organized to maximize their usefulness for both students and teachers. They:

- address fundamental curriculum topics and crucial learning skills, and support essential student learning about science.
- are designed to be visually interesting and academically appropriate. Information is presented in a logical format. In most cases, the graphic organizers are self-explanatory. There is no superfluous information, and there is ample room for students to write in the answer spaces. Form follows function; these graphic organizers direct and enhance student understanding as opposed to clouding it.
- were developed in keeping with the national curriculum standards advocated by the National Science Teachers Association. Every graphic organizer meets at least one, and often several, national and state curriculum standards.

Introduction

To assist you in making the most of these graphic organizers, each one is supported in the "Teaching Notes" section with an objective, teaching tips, and discussion questions. These will help you introduce, teach, and extend each graphic organizer.

The graphic organizers can be used as previews, lessons, reviews, or extension activities. They can be assigned as homework or for in-class completion and done by individual students, student pairs, small groups, or the entire class.

You are encouraged to modify these graphic organizers according to your specific classroom needs. For example, you might direct students to add lines and cells to incorporate additional information, or have them cross out parts of a graphic organizer they do not need.

However you choose to use them, you will find the graphic organizers in *Graphic Organizers: Science* an invaluable addition to your teaching resources.

Research and Standards

Decades of research have consistently demonstrated the value of graphic organizers to both teachers and students. Indeed, few other classroom tools have been so widely and highly praised by education researchers, and so uniformly declared a "best practice."

Education professor Gloria A. Dye (2000) summarized the benefits of graphic organizers:

> Graphic organizers can benefit teachers, students, and parents. By using graphic organizers, teachers can have a clear understanding of what they want to address in their classes. Graphic organizers provide students with a road map to follow as they expand their schemas by linking them to existing knowledge. When looking at a graphic organizer ... parents can have a clearer view of what their children will learn or what they have learned in a content-area class. By serving as a means of helping students link the new information with their existing knowledge base, graphic organizers can contribute to our ultimate goal—student learning.

Specifically, studies have established that graphic organizers:

- work at every grade level;
- are useful at every step of the teaching cycle;
- augment teacher effectiveness;
- apply to students of all ability levels, including those with special needs;
- promote both comprehension and higher-order thinking skills;
- improve decision-making;
- enhance reading comprehension; and
- sharpen research skills.

Meeting National Standards

Graphic Organizers: Science supports the authoritative national education standards put forth by the National Committee on Science Education Standards and Assessment, National Research Council (NRC) in its *National Science Education Standards: Observe, Interact, Change, Learn.* These standards are supported by the National Science Teachers Association. The content standards are organized into eight categories (NRC, 1996):

- Unifying Concepts and Processes in Science
- Science as Inquiry
- Physical Science
- Life Science
- Earth and Space Science
- Science and Technology
- Science in Personal and Social Perspectives
- History and Nature of Science

The structure of *Graphic Organizers: Science* reflects this presentation, and each activity sheet supports a specific topic within a content standard.

Baxendell, B. (2003, January-February). Consistent, coherent, creative: The 3 c's of graphic organizers. *Teaching Exceptional Children, 35,* 46–53.

Dye, G. (2000, January-February). Graphic organizers to the rescue! Helping students link—and remember—information. *Teaching Exceptional Children, 32,* 72–76.

Egan, M. (1999, May). Reflections on effective use of graphic organizers. *Journal of Adolescent & Adult Literacy, 42,* 641–645.

Katayama, A. & Robinson, D. (2000, Winter). Getting students 'partially' involved in note-taking using graphic organizers. *Journal of Experimental Education, 68,* 119–133.

Kealy, W. (2000, Fall). The role of semantic congruency in the design of graphic organizers. *Quarterly Review of Distance Education, 1,* 205–214.

National Research Council. (1996). *National science education standards: Observe, interact, change, learn.* Washington, DC: The National Academies Press.

Teaching Notes

Scientific Inquiry

1. What Is Science? (page 1)

Objective

Students will define science and identify its major characteristics.

Teaching Tips

Use this graphic organizer early in the year to provide students with a context for learning. Consider giving them one example for each box in the graphic organizer. Also, use graphic organizer 46, "Personal Science Glossary," and graphic organizer 47, "K-W-L Chart," as needed.

Discussion Questions

- What is science?
- How is science a special way of understanding the world?
- Which characteristic of science do you think is the most important? Why?

2. Branches of Science (page 2)

Objective

Students will identify and describe the major branches of physical and life sciences.

Teaching Tips

Use this graphic organizer early in the year to provide students with a context for learning. Direct them to name branches where indicated. Also, use graphic organizer 46, "Personal Science Glossary"; graphic organizer 47, "K-W-L Chart"; and graphic organizer 52, "Venn Diagram Template," as needed.

Discussion Questions

- What are the branches of science?
- Why is science organized this way?
- Why is it important to understand this organization?
- How are the branches of science further divided?
- How is each branch of science used?

Teaching Notes

3. The Scientific Method (page 3)

Objective

Students will identify, place in order, and describe the major steps in the scientific method.

Teaching Tips

Make sure students identify the steps in the correct order:

1. Observe carefully.
2. Identify a problem or question.
3. Make a prediction (hypothesis).
4. Design an experiment to test your hypothesis.
5. Conduct an experiment to test your hypothesis.
6. State your findings.

Give further details about each step in a class discussion. Distribute additional copies of this graphic organizer for students to use with particular experiments.

Discussion Questions

- What is the scientific method?
- Why is it important?
- Why is the last step in the scientific method so necessary?
- How do you use versions of the scientific method to make decisions in your everyday life?

4. Tools of Science (page 4)

Objective

Students will identify major tools and equipment used in science.

Teaching Tips

Direct students to write in new tools on this graphic organizer as they learn about and use them. Point out the importance of printed tools. Also emphasize the necessity of using tools safely and properly. Encourage students to include mathematics and logic as basic tools of science in the "Other Important Tools" box.

Discussion Questions

- How is each tool used?
- Which tools seem most interesting or fun to use? Why?

Teaching Notes

5. **Scientific Skills** (page 5)

Objective

Students will identify and describe important scientific skills and perform a self-assessment.

Teaching Tips

Skills that students might list include observing, inferring, communicating, measuring, and reading graphs. Encourage students to update this graphic organizer as they progress in their study of science.

Discussion Questions

- What are some examples of how you have used each skill?
- Why is it important to practice each skill?
- How well do you know each skill?
- What steps can you take to master these skills?

6. **Important Scientific Formulas** (page 6)

Objective

Students will identify and describe important scientific formulas.

Teaching Tips

Encourage students to update this graphic organizer and use it for reference and review. Also, use graphic organizer 48, "Data Table Template," as needed.

Discussion Questions

- What do the letters and symbols in each formula mean?
- When should each formula be used?

Teaching Notes

Physical Science

7. Matter (page 7)

Objective

Students will define and describe matter.

Teaching Tips

Help students develop a firm grasp of the concept of matter. Emphasize its fundamental nature. Challenge students to explain why heat is not considered matter. Also, use graphic organizer 46, "Personal Science Glossary," as needed.

Discussion Questions

- How did you respond to the questions on the graphic organizer?
- What new things have you learned about matter?

8. Periodic Table of the Elements (page 8)

Objective

Students will analyze and explain the periodic table of the elements.

Teaching Tips

Direct students to keep this graphic organizer with their copy of the periodic table as a reference tool. Also, use graphic organizer 46, "Personal Science Glossary," as needed.

Discussion Questions

- What is the periodic table of the elements?
- Why is it important?
- How is it organized?
- What does it tell you?
- How do you read it?
- Why is the table of the elements called "periodic"?

9. Newton's Laws of Motion (page 9)

Objective

Students will identify and describe Newton's laws of motion and explain their significance.

Teaching Tips

Discuss each law individually and in appropriate detail. Encourage students to memorize Newton's laws of motion. Challenge them to explain why these laws are so important both in a historical and scientific context. Also, use graphic organizer 46, "Personal Science Glossary," as needed.

Discussion Questions

- What are Newton's laws of motion?
- How are they expressed mathematically?
- What are some examples of each law?
- Why are these statements called "laws"?

10. Potential Energy and Kinetic Energy (page 10)

Objective

Students will define, compare, and contrast potential energy and kinetic energy.

Teaching Tips

Emphasize the relationship between potential energy and kinetic energy. Also, use graphic organizer 52, "Venn Diagram Template," as needed.

Discussion Questions

- What is potential energy?
- What is kinetic energy?
- How are the two types of energy alike? Different?
- What is an example of potential energy becoming kinetic energy?

Teaching Notes

11. Energy Transfer (page 11)

Objective

Students will describe, compare, and contrast radiation, convection, and conduction.

Teaching Tips

This graphic organizer works well as a partners activity. Direct students to draw symbols to represent each type of energy transfer. Challenge them to identify examples of energy and energy transfer that they currently see. Also, use graphic organizer 46, "Personal Science Glossary," and graphic organizer 53, "Triple Venn Diagram Template," as needed.

Discussion Questions

- What is radiation?
- What is convection?
- What is conduction?
- How are they alike? Different?

12. Waves (page 12)

Objective

Students will identify and describe the major characteristics of waves.

Teaching Tips

This graphic organizer works well as a whole-class activity. Make sure students understand the meaning of *medium* as it refers to waves. Also, use graphic organizer 46, "Personal Science Glossary," as needed.

Discussion Questions

- What are waves?
- What are the parts of waves?
- What are the major types of waves?
- What are some examples of waves?
- What are some other important ideas about waves?
- How is the information in this graphic organizer related to your lab work?

Teaching Notes

13. Major Parts of Waves (page 13)

Objective

Students will identify, define, and describe the major parts of waves.

Teaching Tips

Make sure students understand they must use some terms more than once on the graphic organizer and that one term will not be used (*frequency*). Also, use graphic organizer 46, "Personal Science Glossary," as needed.

Discussion Questions

- What are the major parts of waves?
- What is the definition of each part?
- How are crests and troughs related to other parts of waves?
- What is the type and medium of the waves shown in the graphic organizer?

14. Light and the Visible Spectrum (page 14)

Objective

Students will explain the visible spectrum and memorize the order of its major colors.

Teaching Tips

Present this graphic organizer as a mnemonic device, and suggest students use the acronym "Roy G. Biv" to remember the order of the colors. Also, use graphic organizer 46, "Personal Science Glossary," as needed.

Discussion Questions

- What is a prism?
- How is wavelength related to color?
- What can you do to memorize this information?
- How is the visible spectrum shown by a rainbow?

15. Electricity and Magnetism (page 15)

Objective

Students will define, compare, and contrast electricity and magnetism.

Teaching Tips

Emphasize the relationship between electricity and magnetism. Also, use graphic organizer 46, "Personal Science Glossary," and graphic organizer 52, "Venn Diagram Template," as needed.

Discussion Questions

- What is electricity?
- What is magnetism?
- How are they alike? Different?
- How do people use each one?
- Why are electricity and magnetism often discussed together?

Life Science

16. Traits of Life (page 16)

Objective

Students will identify, describe, and provide examples of the five traits of living organisms.

Teaching Tips

Direct students to list the five traits they believe make something alive in the box on the left. Then, the actual traits should be listed on the tiger. Have students identify the traits in a wide variety of living organisms. Challenge them to explain why a cloud is not considered to be alive.

Discussion Questions

- What five traits do all living organisms have?
- What are some examples of each of these traits?
- How many of these traits must an organism have to be considered alive?
- Why is it important to tell the difference between living and nonliving things?

Teaching Notes

17. Scientific Classification System (page 17)

Objective

Students will identify the scientific classification system used in biology.

Teaching Tips

Have students classify at least one species through all the units of classification, from broadest to most specific. Challenge students to explain why one level of scientific classification has two names.

Discussion Questions

- What are the units of classification?
- What are the larger units? Smaller units?
- How do these units relate to one another?
- Why is every species of life considered a member of each of these units?

18. Kingdoms of Life (page 18)

Objective

Students will identify and describe the kingdoms of life.

Teaching Tips

Have students use this graphic organizer to help them memorize the five kingdoms of life.

Discussion Questions

- What are the five kingdoms of life?
- What are some examples of organisms in each kingdom?
- What kingdoms do certain organisms belong to?
- Why do you think knowledge about life is organized this way?
- Which two kingdoms seem most alike? Most different?
- What determines which kingdom an organism belongs to?
- How does the division of kingdom relate to other divisions in the scientific classification system?

Teaching Notes

19. Ecosystems (page 19)

Objective

Students will define and provide examples of ecosystems and their major parts.

Teaching Tips

This graphic organizer is designed to help students think of ecosystems as having biotic (living) and abiotic (nonliving) elements. Challenge students to explain how the concept of the biosphere relates to the concept of ecosystems. Also, use graphic organizer 46, "Personal Science Glossary," as needed.

Discussion Questions

- What are ecosystems?
- What are some examples of ecosystems?
- What is the difference between biotic and abiotic parts of ecosystems?
- What are some examples of biotic parts? Abiotic?

20. Cells (page 20)

Objective

Students will identify, define, and describe the functions of major parts of a cell.

Teaching Tips

Have students keep this graphic organizer in their notebooks as a quick-reference tool. Emphasize that knowledge of cells is so important because cells are the building blocks of life. Challenge students to explain how the term *organelles* relates to the major parts of a cell.

Discussion Questions

- What are the major parts of a cell?
- What role does each part perform?
- What cell parts are found only in plant cells?
- Which part of a cell "manages" the other parts of a cell?

Teaching Notes

21. Methods of Reproduction (page 21)

Objective

Students will define, describe, compare, and contrast the two main types of reproduction.

Teaching Tips

Challenge students to explain how the process of mitosis is essential to both types of reproduction. Ask them which seem greater: the likenesses or differences between the two types of reproduction, and why. Also, use graphic organizer 46, "Personal Science Glossary," and graphic organizer 52, "Venn Diagram Template," as needed.

Discussion Questions

- What is reproduction?
- Why is it important to life?
- What are the two main types of reproduction?
- How are they alike? Different?
- What are some examples of species that rely on sexual reproduction? Asexual?

Earth and Space Science

22. The Solar System (page 22)

Objective

Students will identify and describe the major parts of the solar system.

Teaching Tips

Have students define *planet.* Instruct them to use the information from this graphic organizer to make a drawing or model of the solar system. Also, use graphic organizer 46, "Personal Science Glossary," and graphic organizer 52, "Venn Diagram Template," as needed.

Discussion Questions

- What is the solar system?
- What are the major parts of the solar system?
- What are the characteristics of each part?
- What is the order of the planets from the sun?
- How are the inner and outer planets alike? Different?

23. Structure of the Earth (page 23)

Objective

Students will identify and describe the internal structure of the earth.

Teaching Tips

Challenge students to speculate on how scientists have learned about the internal structure of the earth. Also, use graphic organizer 46, "Personal Science Glossary," as needed.

Discussion Questions

- How many major layers are there within the earth?
- How thick is each layer?
- What is the thickest layer? Thinnest?
- What is each layer made of?

24. Plate Tectonics (page 24)

Objective

Students will define and explain plate tectonics.

Teaching Tips

This graphic organizer works well as a review sheet or for recording information as it is learned. Have students define *fault* and explain how it relates to plate tectonics. Also, use graphic organizer 46, "Personal Science Glossary," as needed.

Discussion Questions

- What is plate tectonics?
- What are plates?
- How many plates are there?
- Why do the plates move?
- How does plate tectonics cause the creation of mountains? of earthquakes?
- Why is understanding plate tectonics important?

Teaching Notes

25. Rocks and Minerals (page 25)

Objective

Students will define, compare, and contrast rocks and minerals.

Teaching Tips

Challenge students to list ways that people use rocks and minerals. Remind students to use the terms *rock* and *mineral* correctly and not to view them as synonyms. Also, use graphic organizer 46, "Personal Science Glossary," and graphic organizer 52, "Venn Diagram Template," as needed.

Discussion Questions

- What is a rock? A mineral?
- What are the differences between rocks and minerals? Likenesses?
- What are the three types of rocks?
- How does the way the rocks formed relate to how they are categorized?

26. The Atmosphere (page 26)

Objective

Students will identify and describe major characteristics of the earth's atmosphere.

Teaching Tips

Have students identify the layers of the earth's atmosphere. Encourage them to memorize the names and order of the layers. Also, use graphic organizer 46, "Personal Science Glossary," as needed.

Discussion Questions

- What is the atmosphere?
- What makes up the atmosphere?
- What are the layers of the atmosphere?
- Why is the atmosphere so important?

Teaching Notes

27. The Water Cycle (page 27)

Objective

Students will identify, place in order, and describe major steps in the water cycle.

Teaching Tips

Have students identify the state that water is in at each stage of the water cycle. Make sure they understand why this process is called a "cycle." Challenge them to describe what is happening in the water cycle outside the classroom at a particular moment. Also, use graphic organizer 46, "Personal Science Glossary," as needed.

Discussion Questions

- What is another name for the water cycle?
- What are the major stages of the water cycle?
- What role does evaporation play in the water cycle? Condensation? Precipitation? Collection?

28. Severe Weather (page 28)

Objective

Students will identify severe weather and explain how three types of it form.

Teaching Tips

Include discussion of safety rules for severe weather. Explain that it is inevitable, but people can take steps to protect themselves. Have students identify the types of severe weather that might strike their communities and the safety steps they should take. Take care to avoid frightening them. Also, use graphic organizer 46, "Personal Science Glossary," as needed.

Discussion Questions

- What is severe weather?
- What are some examples of severe weather?
- What causes each type of severe weather?

Teaching Notes

Science and Technology

29. What Is Technology? (page 29)

Objective

Students will define and provide examples of technology and explain why it is important.

Teaching Tips

Have students give examples of technology they see around them. Help them understand *technology* as a broad term that covers all of the ways humans use technological means. Also, use graphic organizer 46, "Personal Science Glossary," as needed.

Discussion Questions

- What is technology?
- Why is it important?
- What are some examples of technology?

30. The Difference Between Science and Technology (page 30)

Objective

Students will compare and contrast science and technology.

Teaching Tips

Challenge students to explain why technology is sometimes called "science in action" or "science put to use." Emphasize that the development of new technologies rests on a foundation of scientific knowledge. Also, use graphic organizer 46, "Personal Science Glossary," and graphic organizer 52, "Venn Diagram Template," as needed.

Discussion Questions

- What is the difference between science and technology?
- What are some examples of each?
- How does science lead to technology?

Teaching Notes

31. Effects of a Technology (page 31)

Objective

Students will identify a particular technology as a cause and list its effects.

Teaching Tips

Emphasize that unplanned consequences accompany the introduction of any new technology. Make sure students understand the cause-effect relationship, and point out how effects become causes in turn. Consider having students construct cause-effect diagrams using one of the identified effects from the graphic organizer as a cause.

Discussion Questions

- What have been the effects of this technology?
- Have any of the effects been more important or far-reaching than the others?
- Which effects were predicted? Unpredicted?

32. Costs and Benefits of a Technology (page 32)

Objective

Students will identify the costs and benefits of a particular technology and weigh them against each other.

Teaching Tips

Point out that every technology, no matter how beneficial, requires some costs. Explain that a *cost-benefit analysis* is a study where the costs of a technology are weighed against the benefits to determine whether the technology is worthwhile. Challenge students to discuss whether they think the costs of their chosen technology outweigh the benefits or vice versa, and why. Also, use graphic organizer 46, "Personal Science Glossary," as needed.

Discussion Questions

- What are the disadvantages or costs of this technology?
- What are the advantages or benefits of this technology?
- Do the benefits outweigh the costs? Explain.

Teaching Notes

33. Technologies of Different Cultures (page 33)

Objective

Students will identify and describe the technological contributions of different cultures.

Teaching Tips

Encourage students to explore the technological contributions of different cultures in various times and places. Present a technology and challenge students to identify its cultural origin.

Discussion Questions

- What technological contributions did each culture make?
- Which do you think are the most important? Why?
- Do any of these technologies affect your life today? Explain.
- Why is it important to know about the technological contributions of different cultures?

Personal and Social Perspectives

34. Health (page 34)

Objective

Students will define health and its components and explain the steps they can take to become and remain healthy.

Teaching Tips

Challenge students to explain the statement, "A truly healthy person has a balance among all three kinds of health." Also, use graphic organizer 46, "Personal Science Glossary," and graphic organizer 53, "Triple Venn Diagram Template," as needed.

Discussion Questions

- What is health?
- What are the three kinds of health?
- How is each kind of health alike? Different?
- How can each kind of health lead to other kinds of health?
- Why is health so important?

Teaching Notes

35. Natural Hazards (page 35)

Objective

Students will identify the causes and effects of a natural hazard and suggest ways to minimize or eliminate the hazard and its consequences.

Teaching Tips

This graphic organizer works well as a partners activity. Assign different students or student pairs various natural hazards to investigate. Have students research and identify disastrous examples of each hazard, but avoid frightening them. Explain that natural hazards are inevitable, but people can take steps to protect themselves. Also, use graphic organizer 46, "Personal Science Glossary," as needed.

Discussion Questions

- What is a natural hazard?
- Why might one take place?
- How can people prevent or lessen the chances of one happening?
- What effects can a natural hazard have on people?
- How can people prevent or lessen these effects?

36. What Makes a Good Scientist? (page 36)

Objective

Students will identify and explain the qualities of a good scientist.

Teaching Tips

Have students compare and contrast their responses and develop a class consensus. Emphasize that these qualities are within their reach. Help them identify concrete steps they can take to develop each quality.

Discussion Questions

- What qualities should a good scientist have?
- Why is each quality important?
- What can you do to develop each quality?
- Is it important for scientists to have a particular quality? Why or why not?

Teaching Notes

37. Profile of a Job in Science (page 37)

Objective

Students will investigate jobs in science and share what they learn.

Teaching Tips

Assign different jobs to various students. Address a wide range of jobs (different levels of education required, numerous fields and workplaces, etc.). Consider assembling student graphic organizers into a "Jobs in Science" booklet or bulletin board display. Invite a person working in a scientific field to visit the class.

Discussion Questions

- What are the major tasks and responsibilities of this job?
- What personal qualities does someone need to be successful at this job?

38. Profile of a Job in Technology (page 38)

Objective

Students will investigate jobs in technology and share what they learn.

Teaching Tips

Assign different jobs to various students. Address a wide range of jobs (different levels of education required, numerous fields and workplaces, etc.). Consider assembling student graphic organizers into a "Jobs in Technology" booklet or bulletin board display. Invite a person working in a technical field to visit the class.

Discussion Questions

- What are the major tasks and responsibilities of this job?
- What personal qualities does someone need to be successful at this job?

Teaching Notes

History and Nature of Science

39. Biographical Profile of a Scientist (page 39)

Objective

Students will conduct research to create a biographical profile of a scientist.

Teaching Tips

Use this graphic organizer to expand on brief references to scientists in textbooks. Consider having students write letters to their profiled scientists, if possible.

Discussion Questions

- What important contributions did this scientist make?
- What events in this person's life might have led him or her to make these contributions?
- Why is it important that we know about this scientist?

40. Profile of a Scientific Milestone (page 40)

Objective

Students will investigate and answer basic questions about a scientific milestone.

Teaching Tips

Allow students to investigate any appropriate scientific event. Challenge them to identify how the scientific milestone affects their lives today.

Discussion Questions

- What is a scientific milestone?
- What made this event a scientific milestone?
- How did you respond to the other questions on the graphic organizer?

Teaching Notes

41. History of a Scientific Development (page 41)

Objective

Students will investigate the cause-effect chain that led to a particular scientific discovery or breakthrough.

Teaching Tips

Make sure students work backwards from the top of the graphic organizer on down. You may wish to provide the following example. The scientific discovery could be, "the discovery of microorganisms." Before microorganisms were discovered, the microscope had to be invented. But before the microscope was invented, lenses had to be understood. But before lenses were understood, methods of making lenses had to be developed. But before these methods were developed, people had to discover how to make glass. Allow students to investigate any appropriate scientific discovery. Emphasize that scientific knowledge is all connected. Challenge students to explain what Isaac Newton meant by, "If I have seen further, it is by standing on the shoulders of giants."

Discussion Questions

- How do scientific discoveries and breakthroughs depend on previous knowledge?
- Is all of this previous knowledge based in science?
- What can you conclude from this?

42. The Scientific Revolution I (page 42)

Objective

Students will identify the Scientific Revolution and discuss its importance.

Teaching Tips

Present the Scientific Revolution as the birth of modern science, and emphasize the fundamental and worldwide changes that this event caused. Also, use graphic organizer 46, "Personal Science Glossary," as needed.

Discussion Questions

- What was the Scientific Revolution?
- Who were some major scientists during that time?
- How did you respond to the other questions on the graphic organizer?

Teaching Notes

43. The Scientific Revolution II (page 43)

Objective

Students will define the Scientific Revolution and identify its key elements.

Teaching Tips

Emphasize the significance of the Scientific Revolution. Engage the class in a debate on whether or not it might qualify as the most influential event in human history. Guide students in creating general descriptions of the world before and after the Scientific Revolution. Also, use graphic organizer 46, "Personal Science Glossary," as needed.

Discussion Questions

- When did the Scientific Revolution happen?
- How has the Scientific Revolution changed people's lives?

44. A Scientific Current Event (page 44)

Objective

Students will recognize that scientific issues are part of the news and will explore a scientific current event in detail.

Teaching Tips

After the students complete the activity, consider compiling the graphic organizers into a class book about scientific news.

Discussion Questions

- What is the event?
- Why is it important?
- How does it relate to your study of science? Your life in general?

Teaching Notes

45. One Week in Science (page 45)

Objective

Students will identify and summarize the major scientific news headlines during one school week.

Teaching Tips

Use this graphic organizer to emphasize the broadness and rapid advancement of science in today's world. Direct students to review the major scientific headlines reported in the press for one school week. Challenge them to identify the branch(es) of science that relate to each headline.

Discussion Questions

- What scientific news was reported?
- Why was each happening important?

Additional Graphic Organizers and Templates

46. Personal Science Glossary (page 46)
Use throughout the instruction period as a way for students to review and remember key terms. Make sure students understand that a *glossary* is a list of specialized terms and their meanings.

47. K-W-L Chart (page 47)
Use as needed. Explain that K-W-L stands for "What I already **K**now," "What I **W**ant to learn," and "What I did **L**earn."

48. Data Table Template (page 48)
Use as needed.

49. Bar Graph Template (page 49)
Use as needed.

50. Line Graph Template (page 50)
Use as needed.

51. Circle Graph Template (page 51)
Use as needed. There are 100 spaces on the circumference.

52. Venn Diagram Template (page 52)
Use as needed. Make sure students understand the reason for the overlapping circles.

53. Triple Venn Diagram Template (page 53)
Use as needed. Make sure students understand the reason for the overlapping circles.

NAME DATE

1. What Is Science?

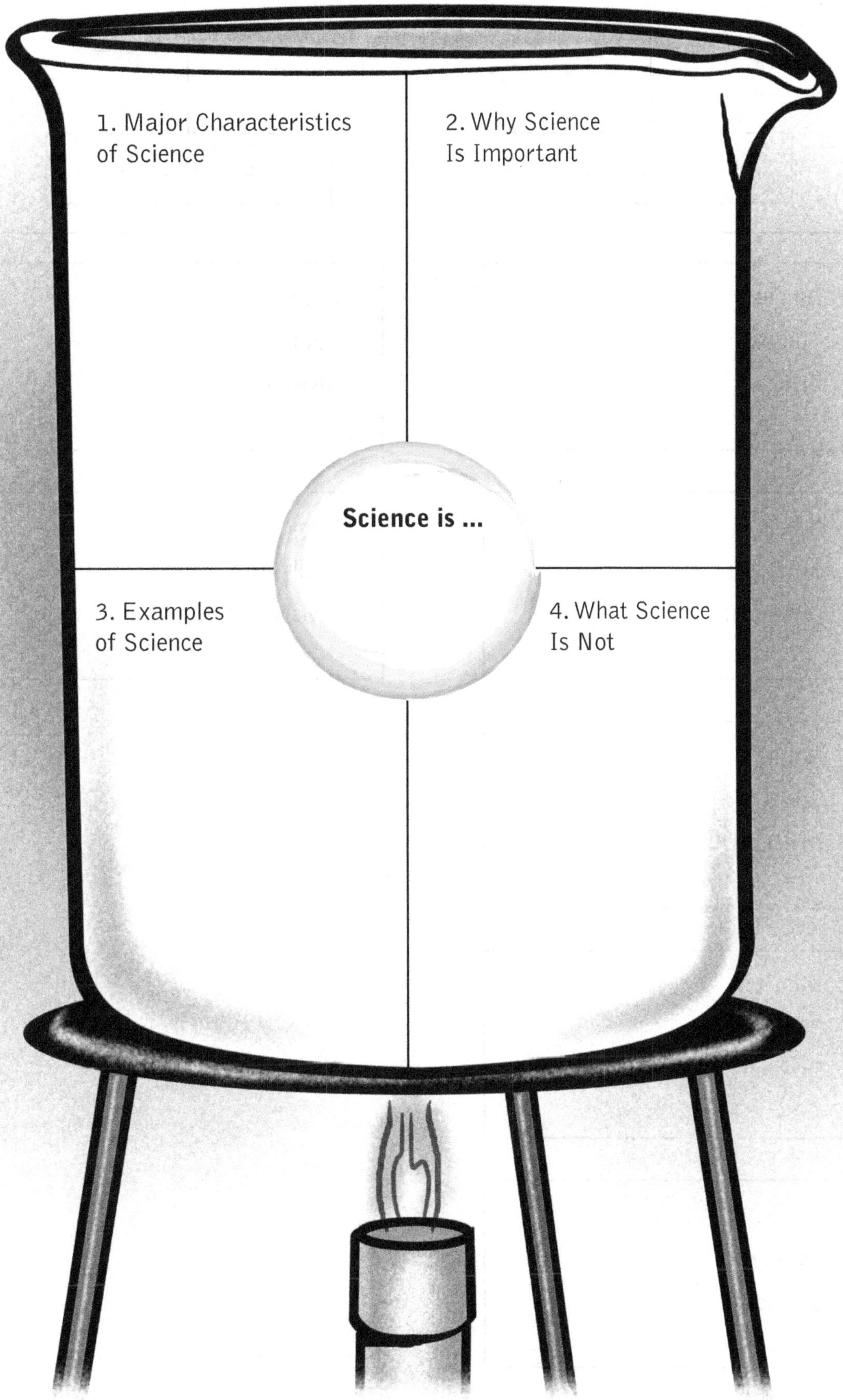

NAME DATE

2. Branches of Science

Science

Physical Sciences

Name:
Study of:
Divisions:

Name:
Study of:
Divisions:

Name:
Study of:
Divisions:

Name:
Study of:
Divisions:

Name:
Study of:
Divisions:

Life Sciences

Name:
Study of:
Divisions:

Name:
Study of:
Divisions:

NAME DATE

3. The Scientific Method

Basic Steps in the Scientific Method

Conduct an experiment to test your hypothesis.	Identify a problem or question.
Design an experiment to test your hypothesis.	Observe carefully.
Make a prediction (hypothesis).	State your findings.

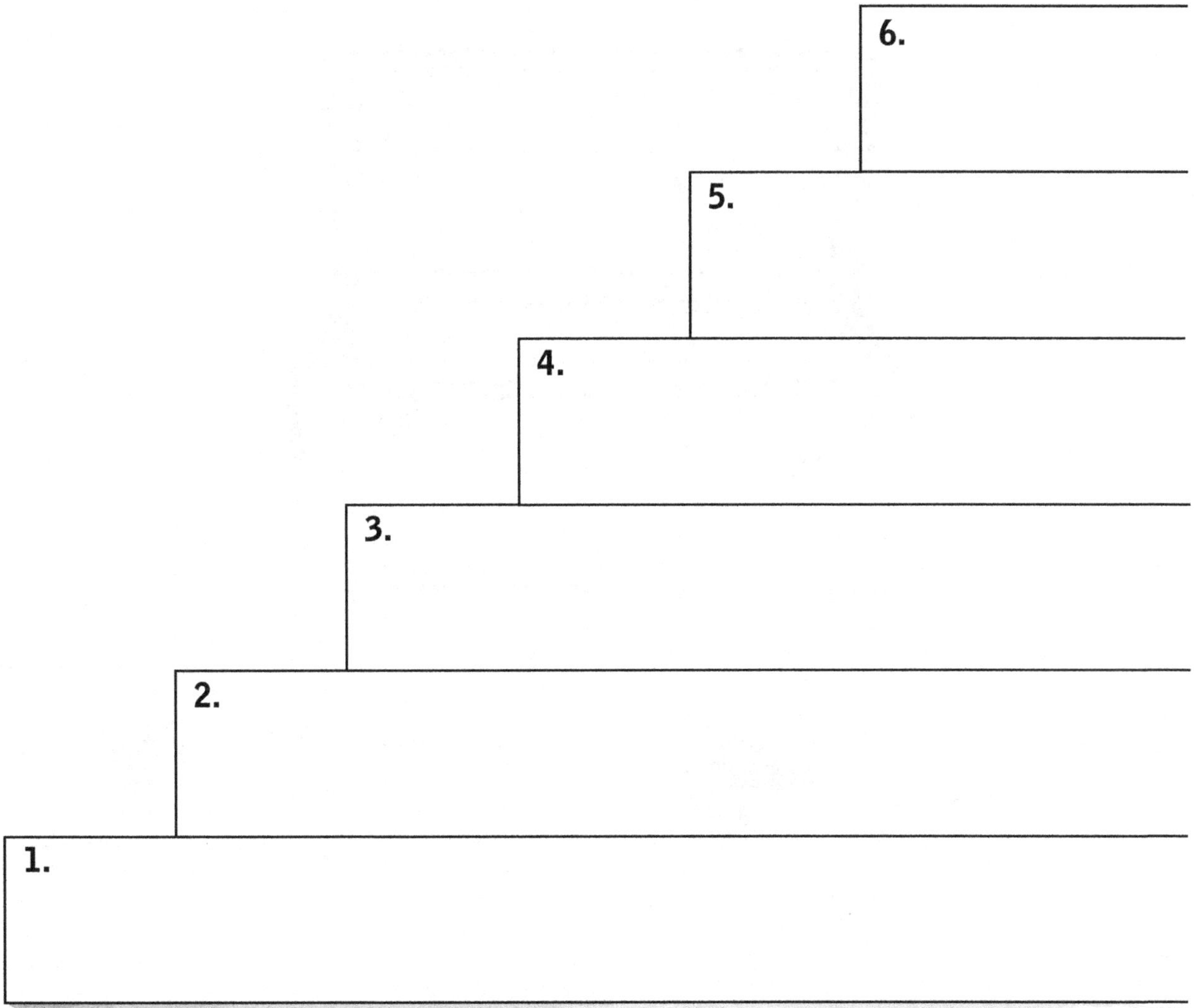

NAME DATE

4. Tools of Science

Laboratory Tools

Printed Tools

Field Tools

Other Important Tools

NAME DATE

5. Scientific Skills

Skill	Description of skill	Why this skill is important	How well do I know this skill? What do I need to do to master this skill?

NAME DATE

6. Important Scientific Formulas

Formula	Purpose and Use	Mathematical Expression	Definition of Terms	Additional Information

NAME DATE

7. Matter

What is matter?

What are three types of matter?

Matter

What are some examples of matter?

What are some things that exist that are not matter?

What is matter made of?

NAME DATE

8. Periodic Table of the Elements

Periodic Table of the Elements

What It Is

Why It Is Important

How It Is Organized

What It Tells You

How to Read It

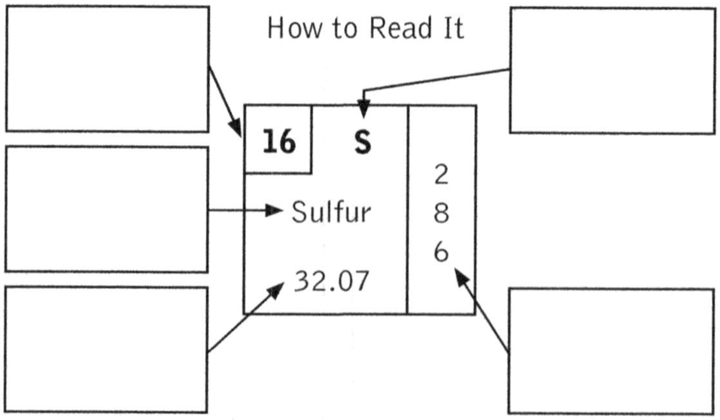

NAME DATE

9. Newton's Laws of Motion

Law	Definition	Mathematical Expression/ Additional Information	Example
Newton's First Law			
Newton's Second Law			
Newton's Third Law			

NAME DATE

10. Potential Energy and Kinetic Energy

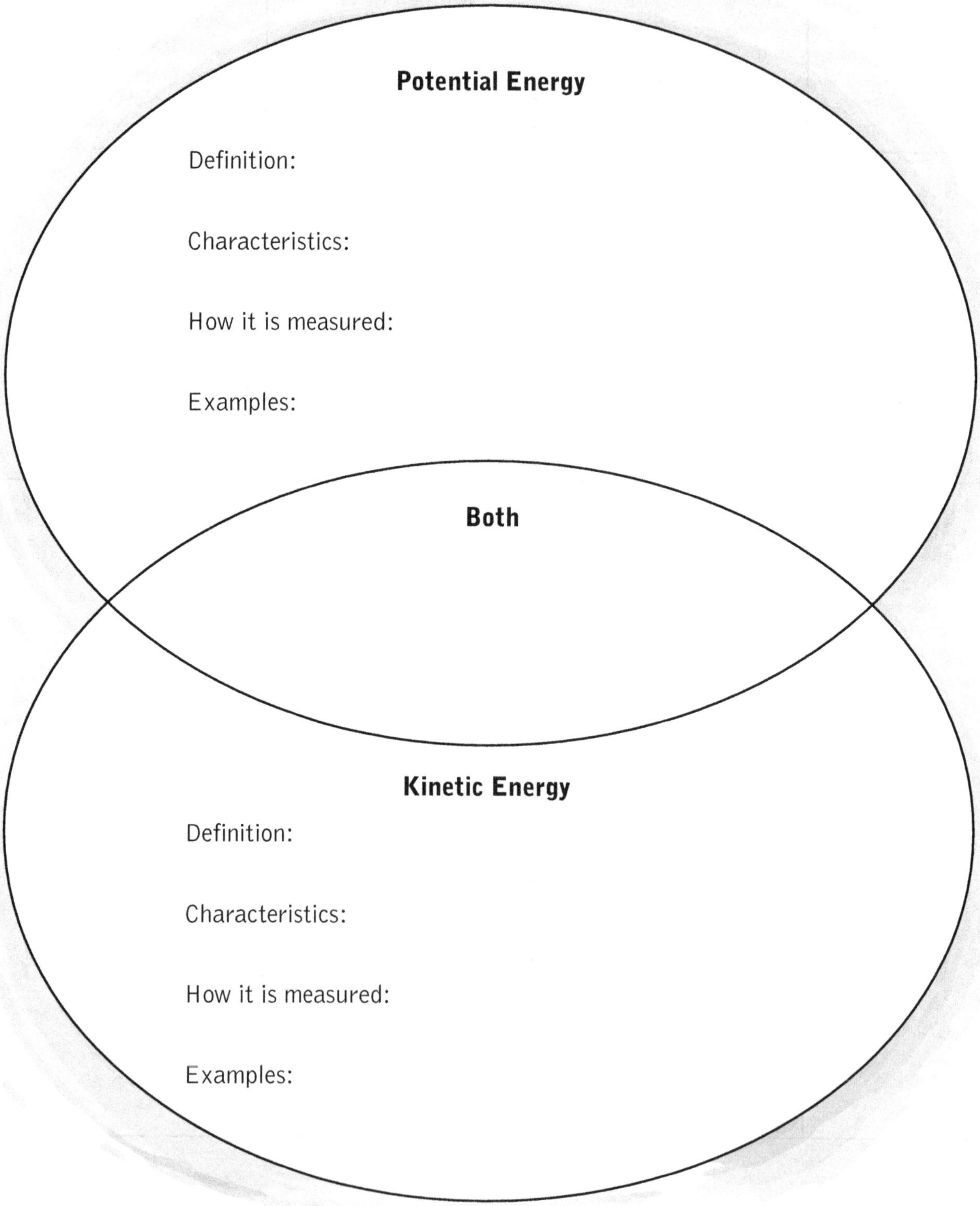

NAME DATE

11. Energy Transfer

Radiation

Description:

Examples:

Convection

Description:

Examples:

The Three Ways Thermal Energy Moves

Conduction

Description:

Examples:

NAME DATE

12. Waves

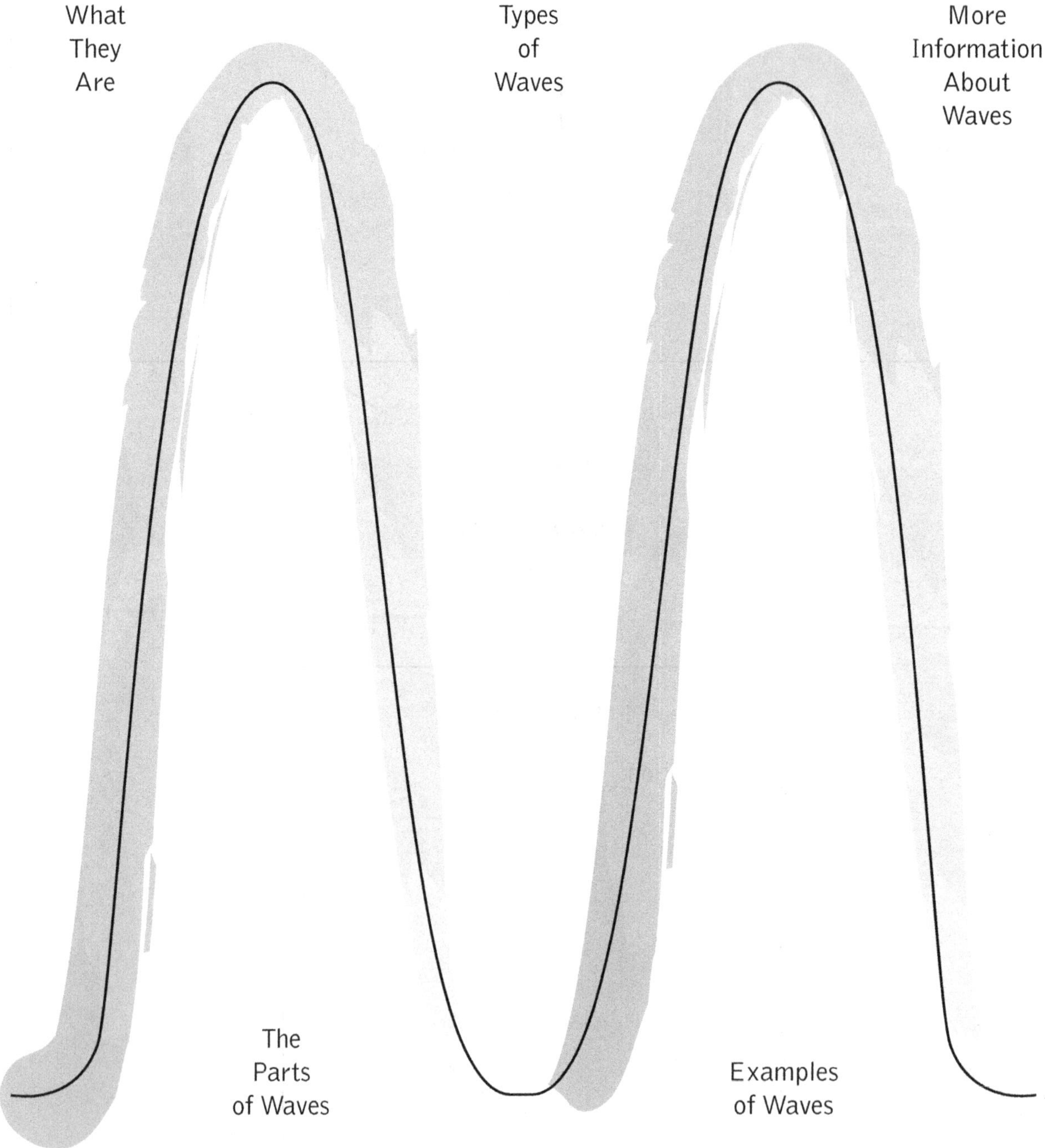

NAME DATE

13. Major Parts of Waves

Waves: Five Important Terms

crest: the top of a wave.
trough: the bottom of a wave.
wavelength: the distance between similar points (like troughs or crests) on neighboring waves.
amplitude: the height of a crest or the depth of a trough above or below the normal level of the medium.
frequency: the number of waves that pass a point in a given amount of time.

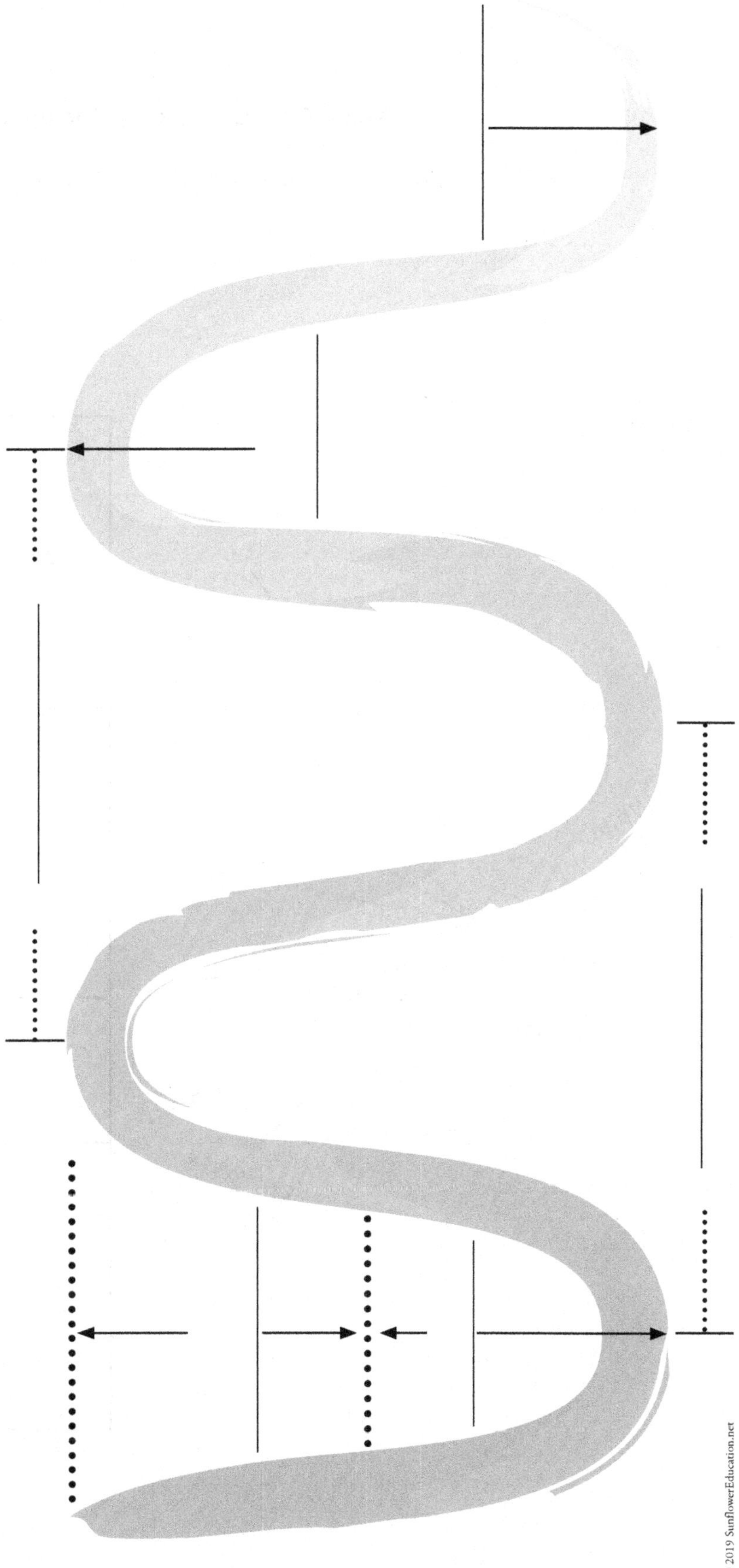

NAME DATE

14. Light and the Visible Spectrum

Major Colors of the Visible Spectrum

red violet orange green yellow blue indigo

Longer Wavelength

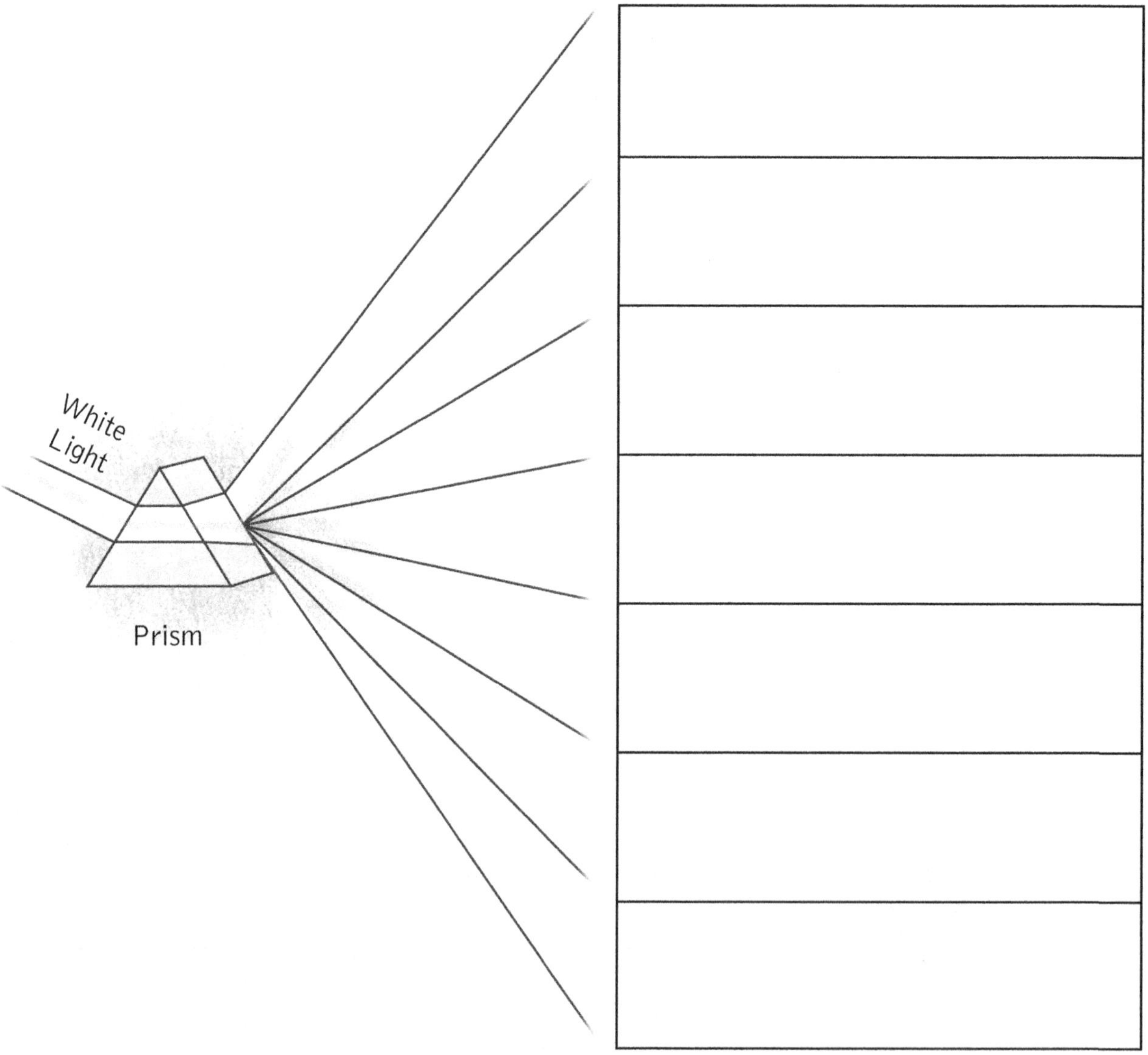

Shorter Wavelength

NAME DATE

15. Electricity and Magnetism

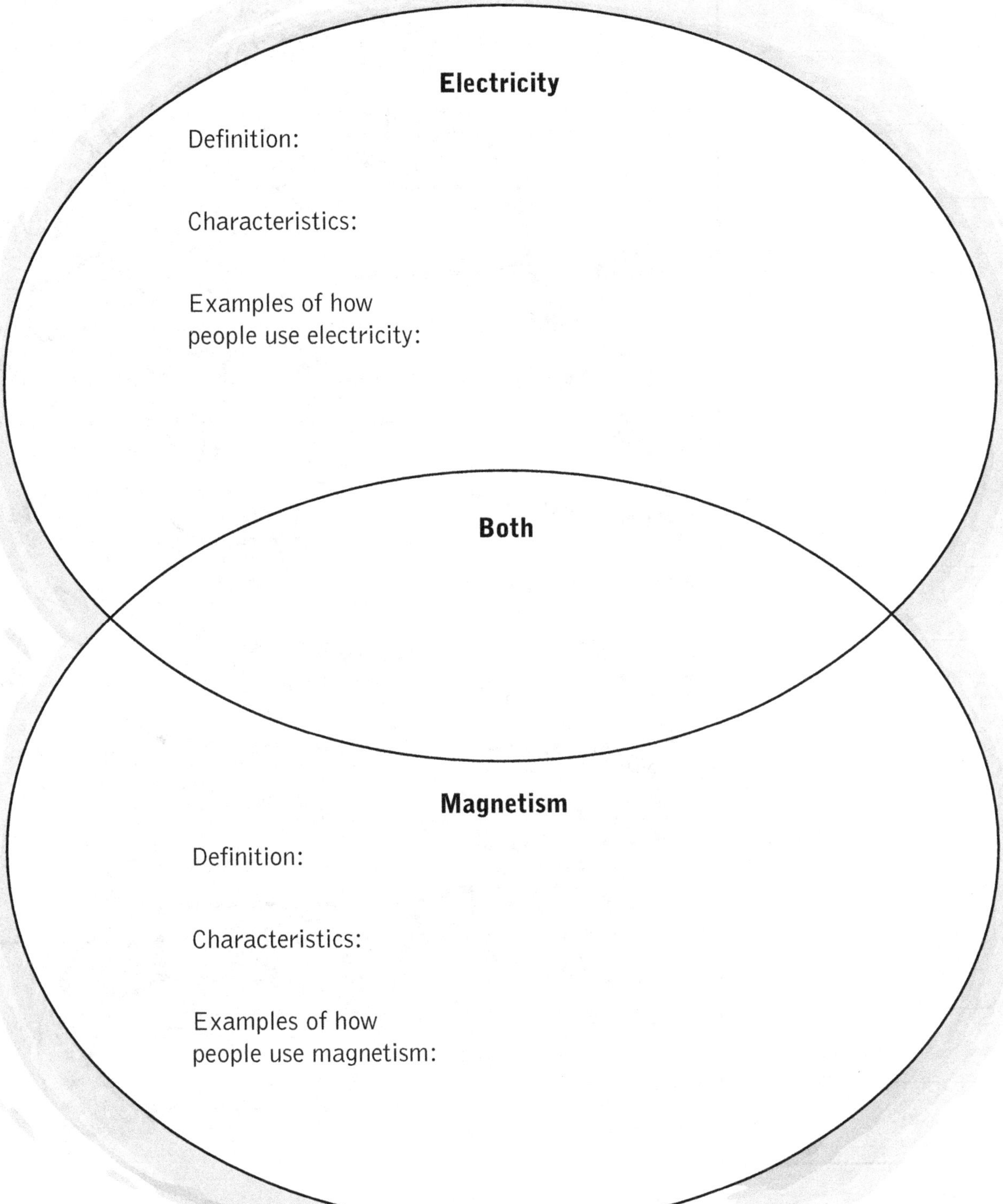

NAME DATE

16. Traits of Life

I Think These Five Traits Make Something Alive

1. ____________________
____
____

2. ____________________
____
____

3. ____________________
____
____

4. ____________________
____
____

5. ____________________
____
____

NAME DATE

17. Scientific Classification System

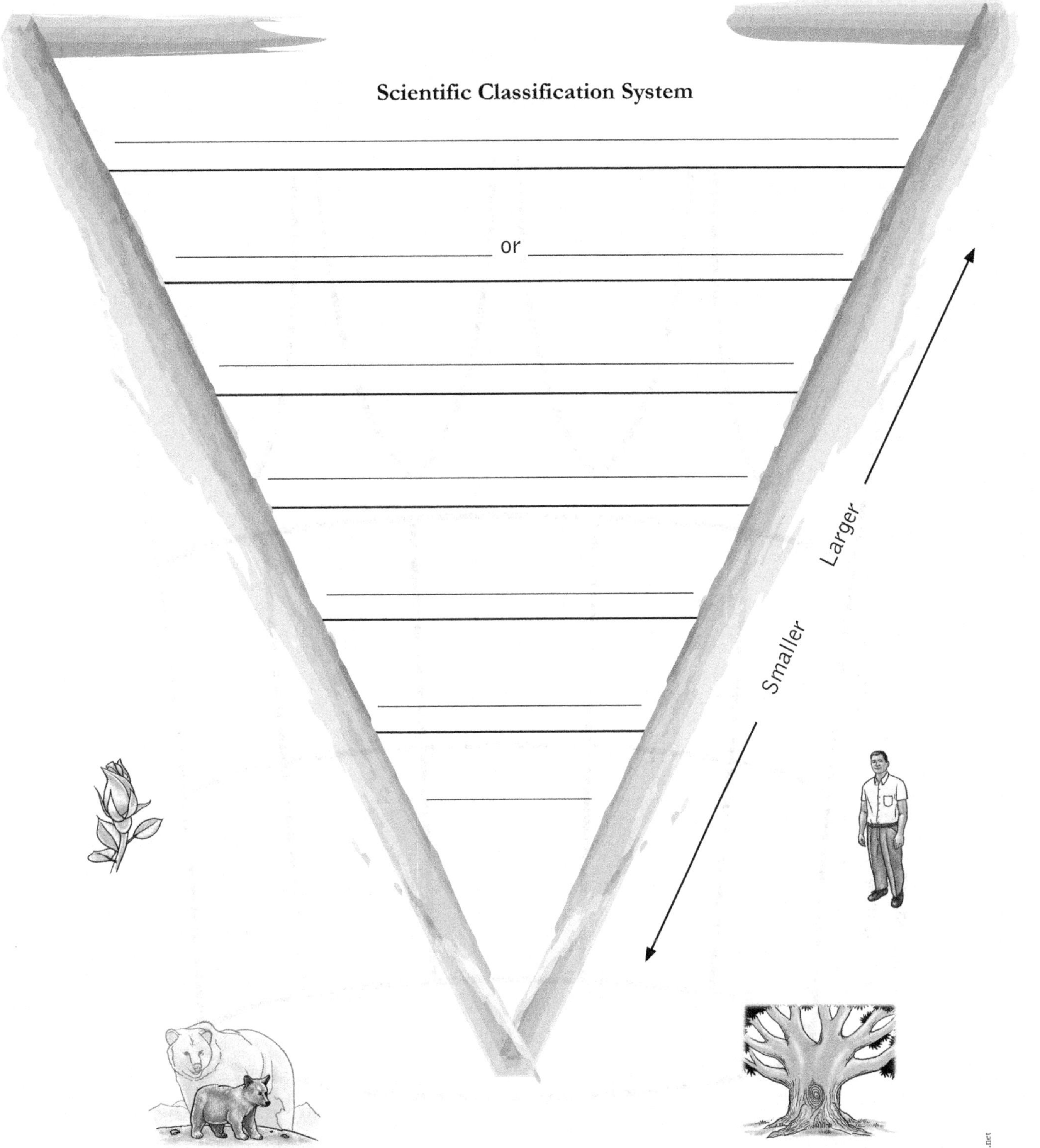

NAME DATE

18. Kingdoms of Life

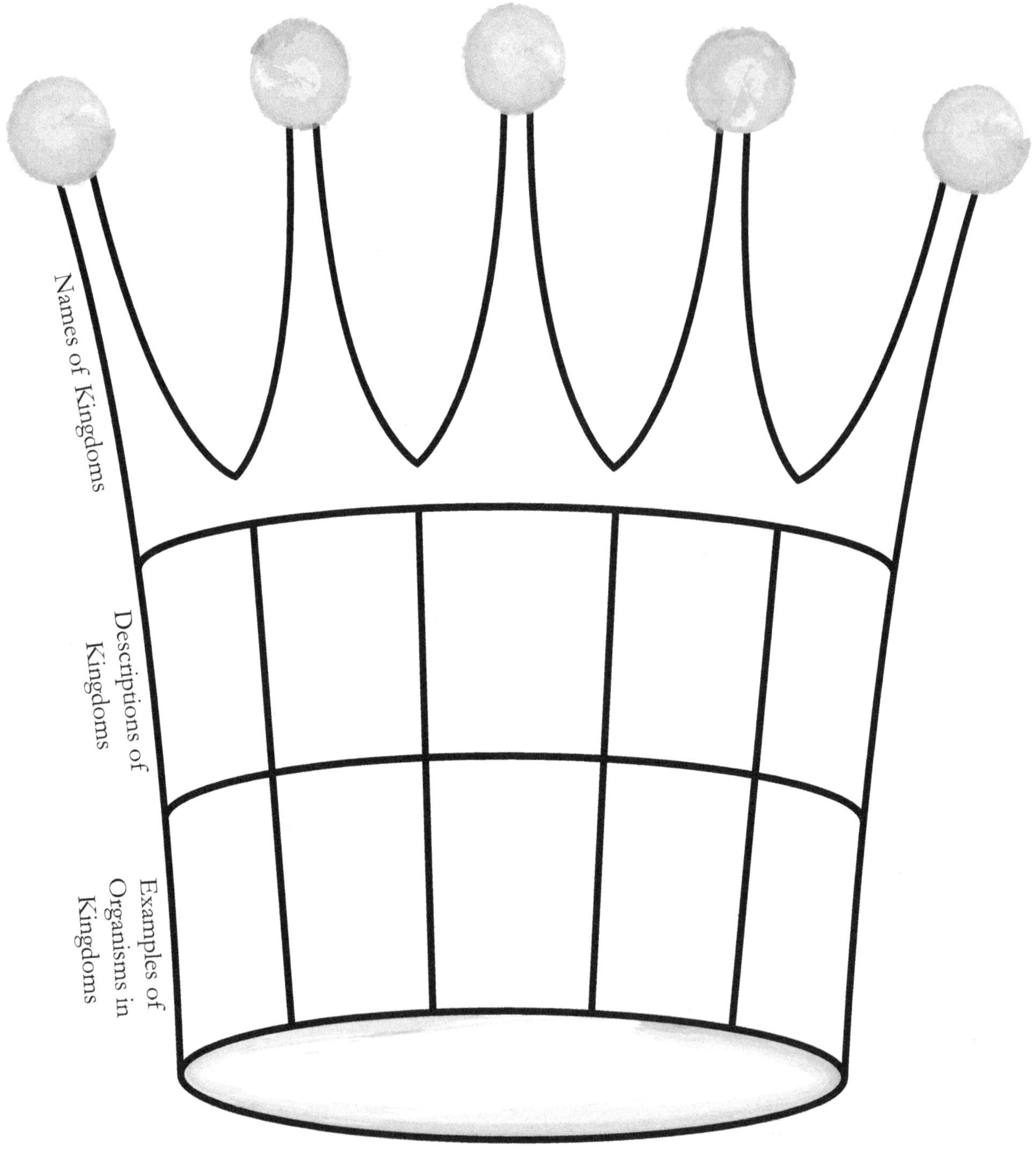

NAME DATE

19. Ecosystems

Ecosystems

are

Examples include

All ecosystems have these two types of factors

and

Examples are

Examples are

NAME DATE

20. Cells

Major Parts of a Cell	
Part	Definition and Description of Function

NAME DATE

21. Methods of Reproduction

Sexual Reproduction

Definition and description:

Examples of organisms that reproduce sexually:

Both

Asexual Reproduction

Definition and description:

Examples of organisms that reproduce asexually:

NAME DATE

22. The Solar System

The Sun			
Inner Planets	General characteristics:	Planet: Characteristics:	Distance from sun:
		Planet: Characteristics:	Distance from sun:
		Planet: Characteristics:	Distance from sun:
		Planet: Characteristics:	Distance from sun:
Asteroid Belt	Definition:	Description:	
Outer Planets	General characteristics:	Planet: Characteristics:	Distance from sun:
		Planet: Characteristics:	Distance from sun:
		Planet: Characteristics:	Distance from sun:
		Planet: Characteristics:	Distance from sun:
		Planet: Characteristics:	Distance from sun:
Comets	Definition:	Description:	

NAME DATE

23. Structure of the Earth

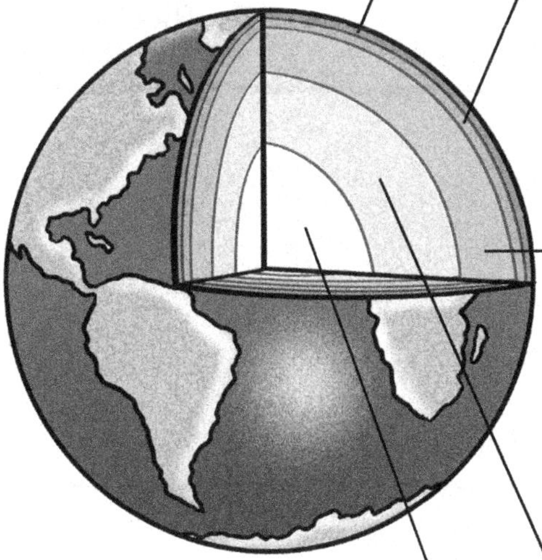

Name:
Thickness:
Description:

Name:
Thickness:
Description:

Name:
Thickness:
Description:

Name:
Thickness:
Description:

Name:
Thickness:
Description:

Deeper

NAME DATE

24. Plate Tectonics

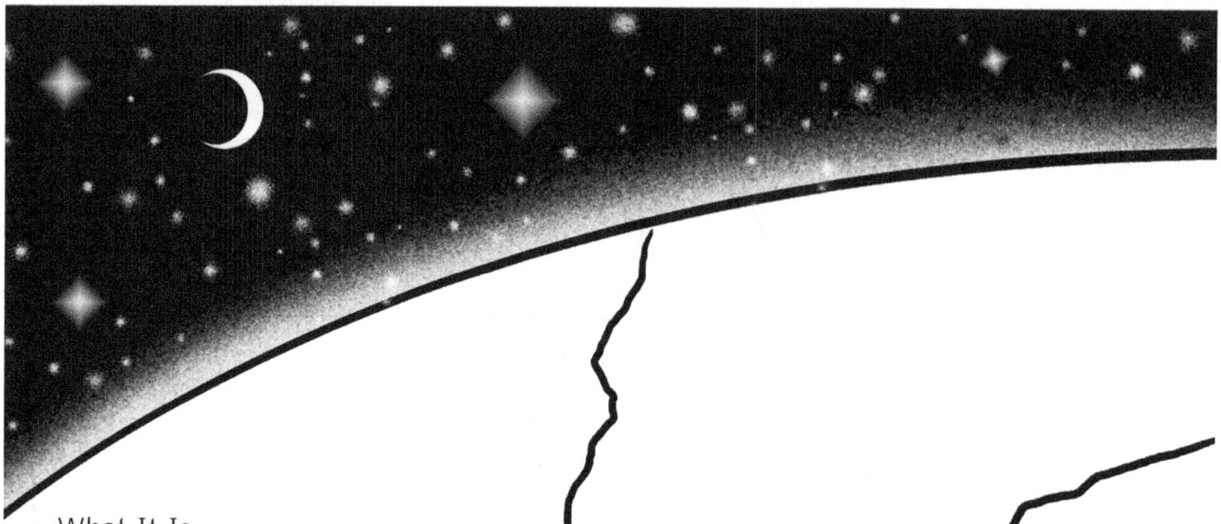

What It Is

What Plates Are

Number of Plates

Why Plates Move

How Plate Movement Changes the Earth

How It Makes Mountains

The Role of Seafloor Spreading

How It Causes Earthquakes

The Role of the Mantle

NAME DATE

25. Rocks and Minerals

Rocks and Minerals

Rocks

Definition:

Three Types of Rocks

Type:
Definition:

Kinds:
Examples:

Type:
Definition:

Kinds:
Examples:

Type:
Definition:

Kinds:
Examples:

Minerals

Definition:

Properties:

Examples:

NAME DATE

26. The Atmosphere

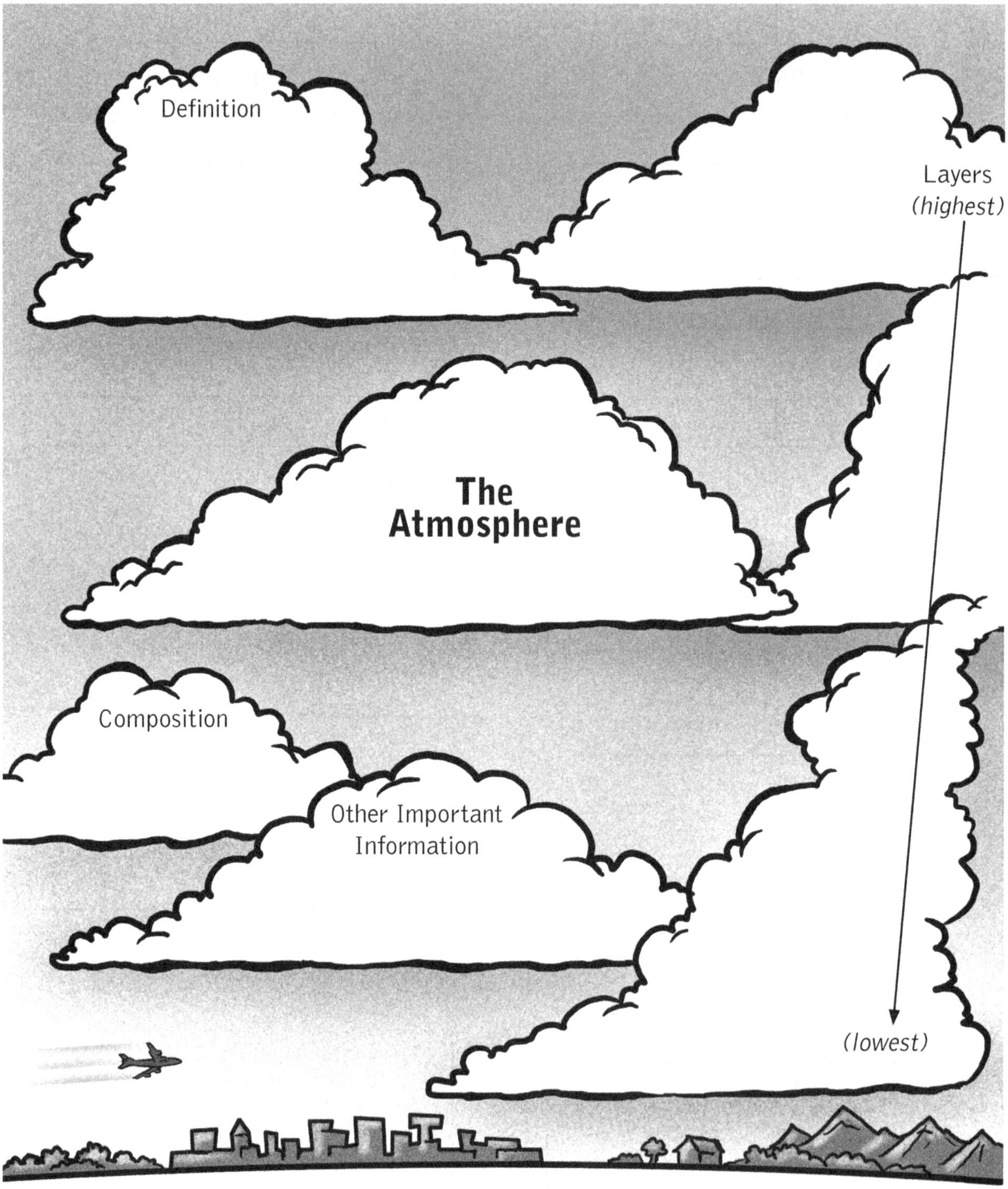

NAME DATE

27. The Water Cycle

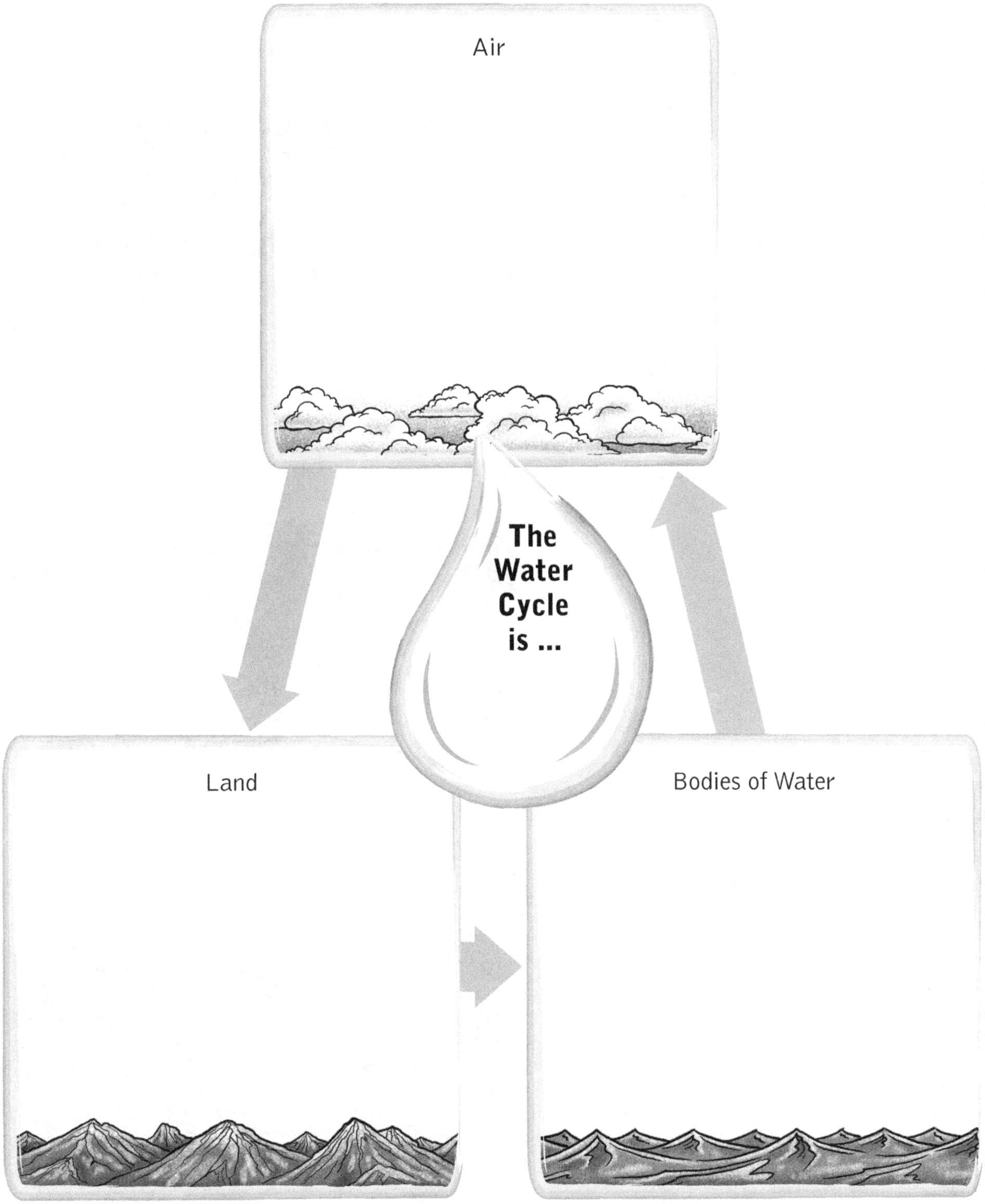

NAME DATE

28. Severe Weather

Causes of Severe Weather

causes

causes

causes

NAME DATE

29. What Is Technology?

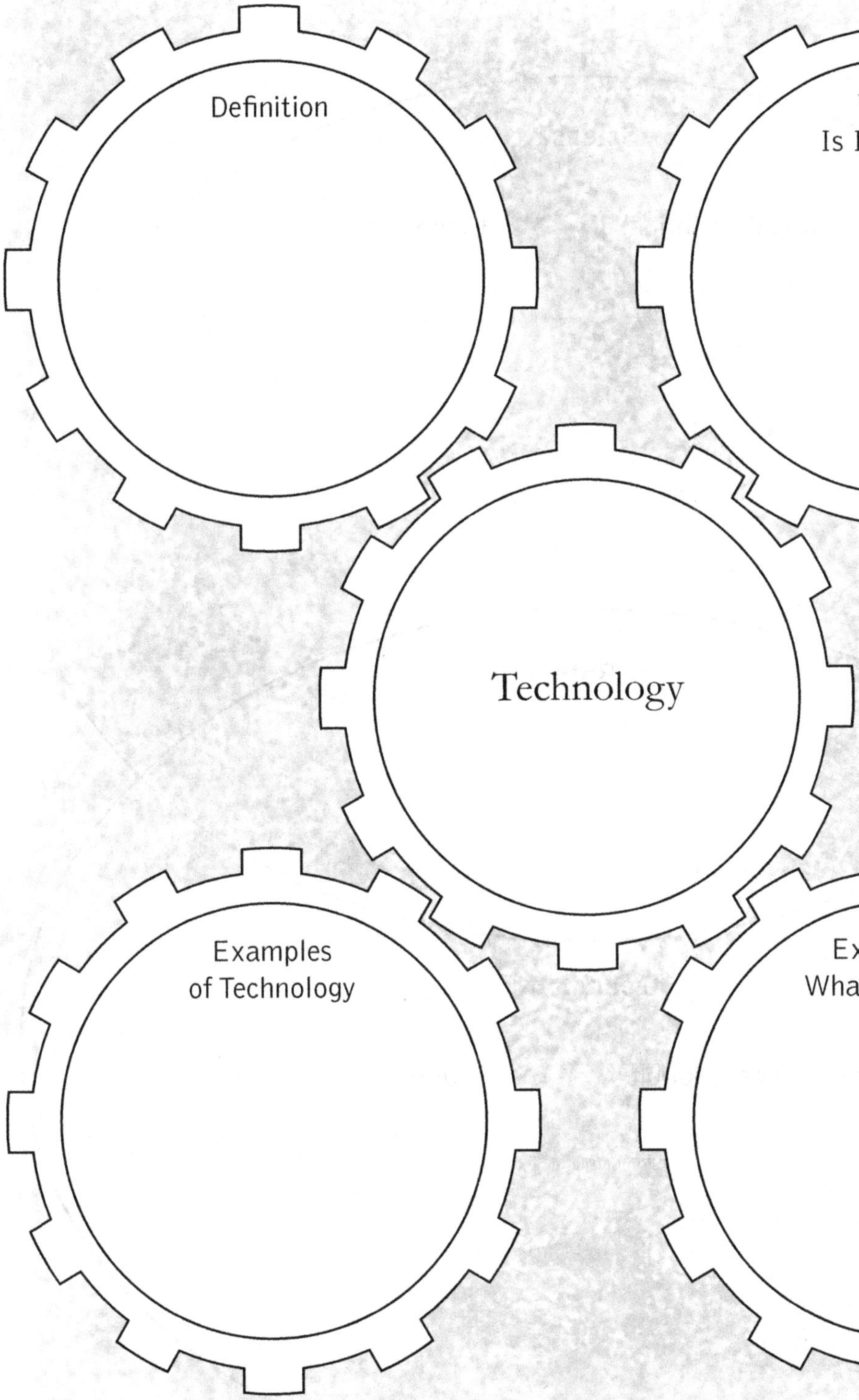

NAME DATE

30. The Difference Between Science and Technology

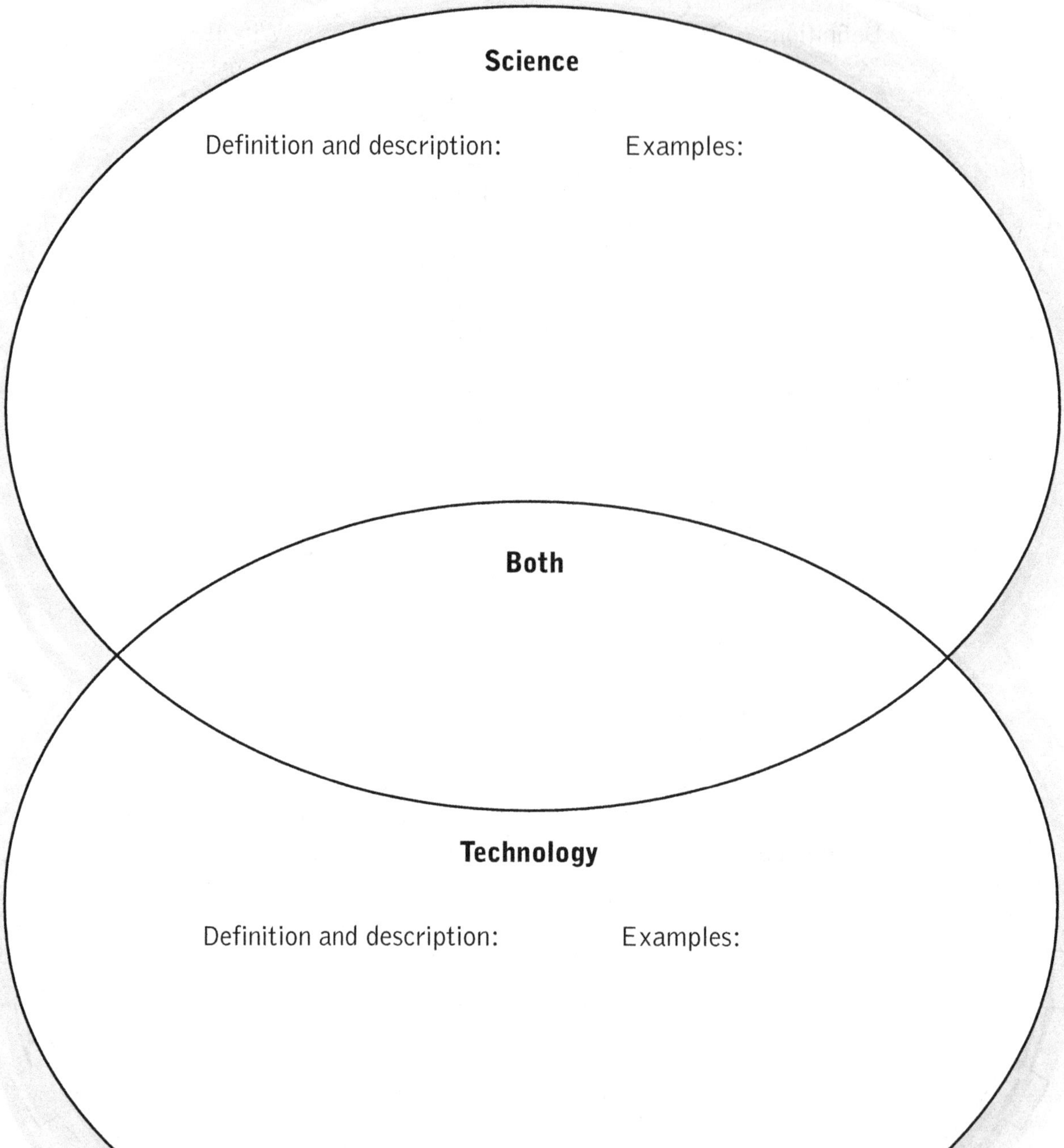

NAME DATE

31. Effects of a Technology

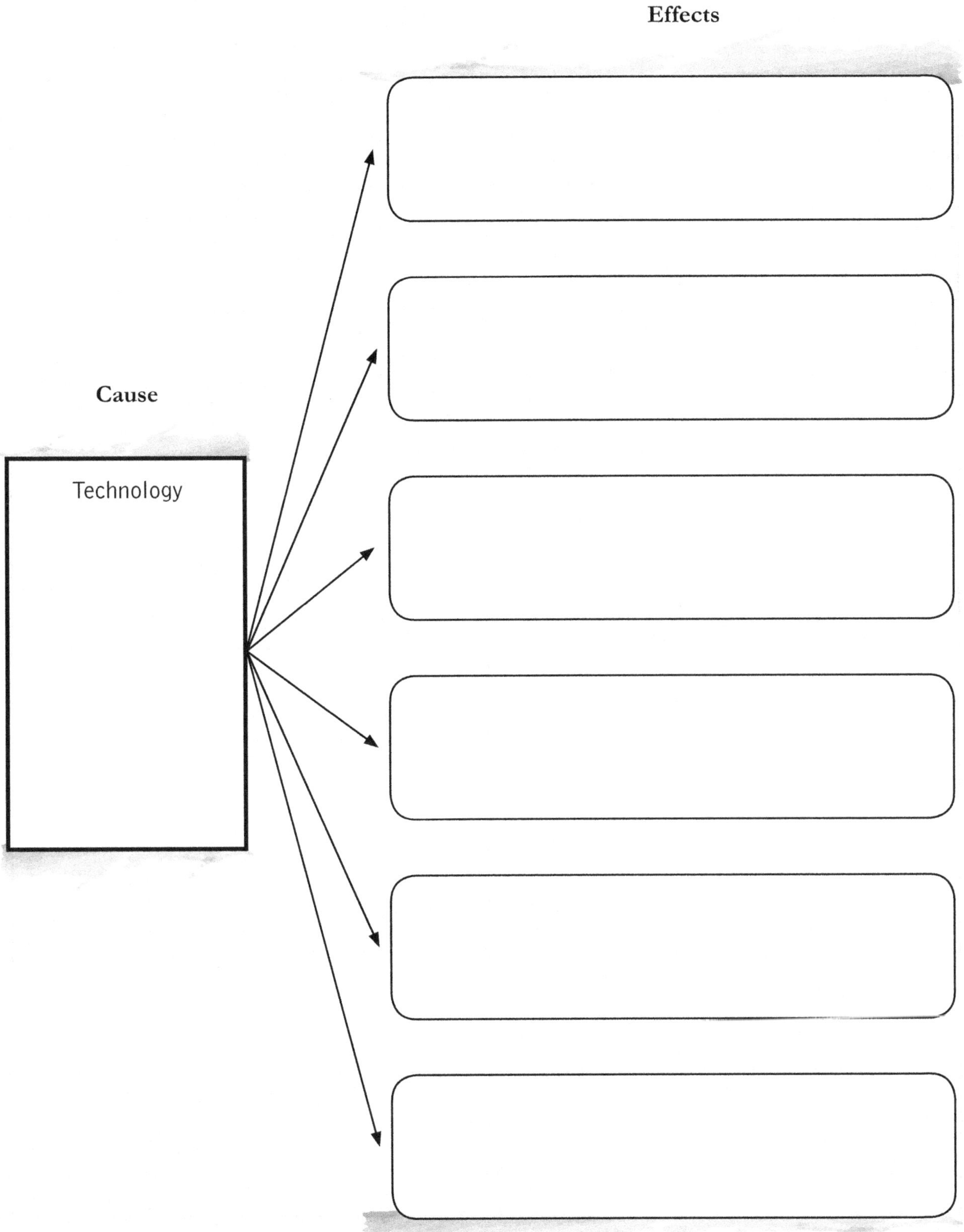

NAME DATE

32. Costs and Benefits of a Technology

Technology: ______	
Costs (-)	Benefits (+)

NAME DATE

33. Technologies of Different Cultures

Technology	Cultural Origin	Approximate Date It Was Introduced

NAME DATE

34. Health

Physical health is ____________

____

____

To be physically healthy, I should

____

____

____

____

Mental health is ____________

____

____

To be mentally healthy, I should

____

____

____

____

Health is ____________

____

____

____

____

____

____

Social health is ____________

____

____

To be socially healthy, I should

____

____

____

____

NAME ______________________ DATE ______________________

35. Natural Hazards

Natural Hazard: ______________________

What causes it

What effects it has on people

What people can do to prevent or lessen the chance of the natural hazard happening

What people can do to prevent or lessen these effects

NAME DATE

36. What Makes a Good Scientist?

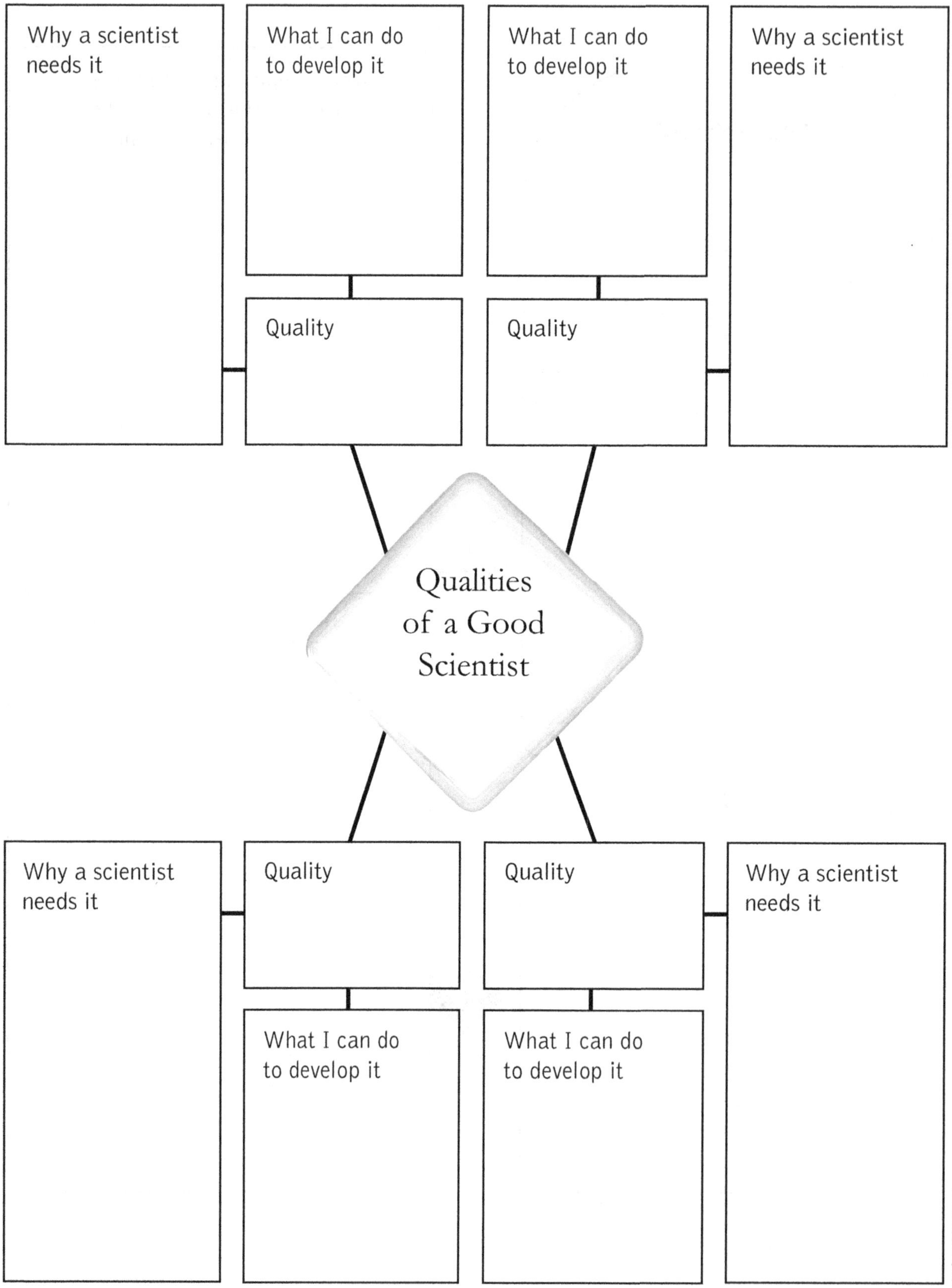

NAME DATE

37. Profile of a Job in Science

A Job in Science	
Name of job	
Name of field this job is in	
Where people doing this job work	
Major tasks and responsibilities of this job	
Why this job is important	
What I would like about doing this job	
What I would not like about doing this job	
Additional information	

NAME DATE

38. Profile of a Job in Technology

A Job in Technology	
Name of job	
Name of field this job is in	
Where people doing this job work	
Major tasks and responsibilities of this job	
Why this job is important	
What I would like about doing this job	
What I would not like about doing this job	
Additional information	

NAME DATE

39. Biographical Profile of a Scientist

The Life and Work of a Scientist	
Name	
Life dates	
Where scientist lived and worked (country, city, other)	
Major fields of study	
Important contributions	
Interesting personal information	
Additional information	

NAME DATE

40. Profile of a Scientific Milestone

A Scientific Milestone

What was it?

Why is it important?

When did it happen?

Where did it happen?

Who made it happen?

How did the individual or group achieve it?

NAME DATE

41. History of a Scientific Development

Scientific Discovery or Breakthrough:

____

But before this happened, what had to occur?

____

But before this happened, what had to occur?

____

But before this happened, what had to occur?

____

But before this happened, what had to occur?

____

But before this happened, what had to occur?

____

But before this happened, what had to occur?

____

NAME DATE

42. The Scientific Revolution I

The Scientific Revolution

What was it?

Why is it important?

When did it happen?

Where did it happen?

Who made it happen?

How did it change the world?

NAME DATE

43. The Scientific Revolution II

The Scientific Revolution

Definition:

When and where it happened:

Leading scientists of the revolution:

The World Before

The World After

NAME DATE

44. A Scientific Current Event

What event is

Where it happened

How I learned of it

A Current Event in Science

Who was involved

When it happened

Why it is important

NAME DATE

45. One Week in Science

One Week in Science	
Day	Major Science Headlines
Monday	• • •
Tuesday	• • •
Wednesday	• • •
Thursday	• • •
Friday	• • •

NAME DATE

46. Personal Science Glossary

Terms Related to My Study of Science	
Term	Definition

NAME DATE

47. K-W-L Chart

Topic: ______		
What I already Know	What I Want to learn	What I did Learn

NAME DATE

48. Data Table Template

____

(title)

NAME DATE

49. Bar Graph Template

(title)

Key

NAME DATE

50. Line Graph Template

(title)

Key

NAME DATE

51. Circle Graph Template

(title)

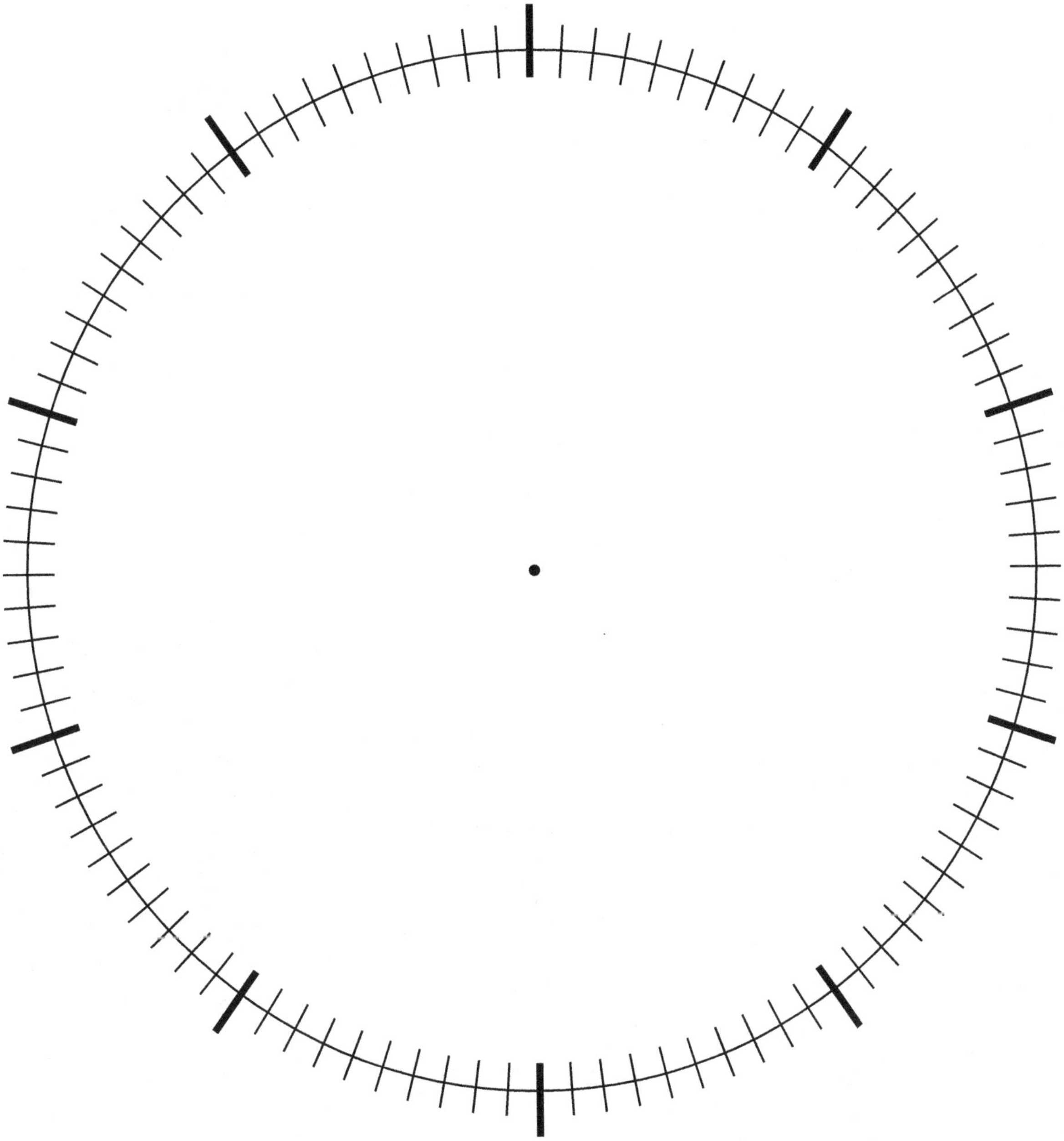

NAME DATE

52. Venn Diagram Template

____

(title)

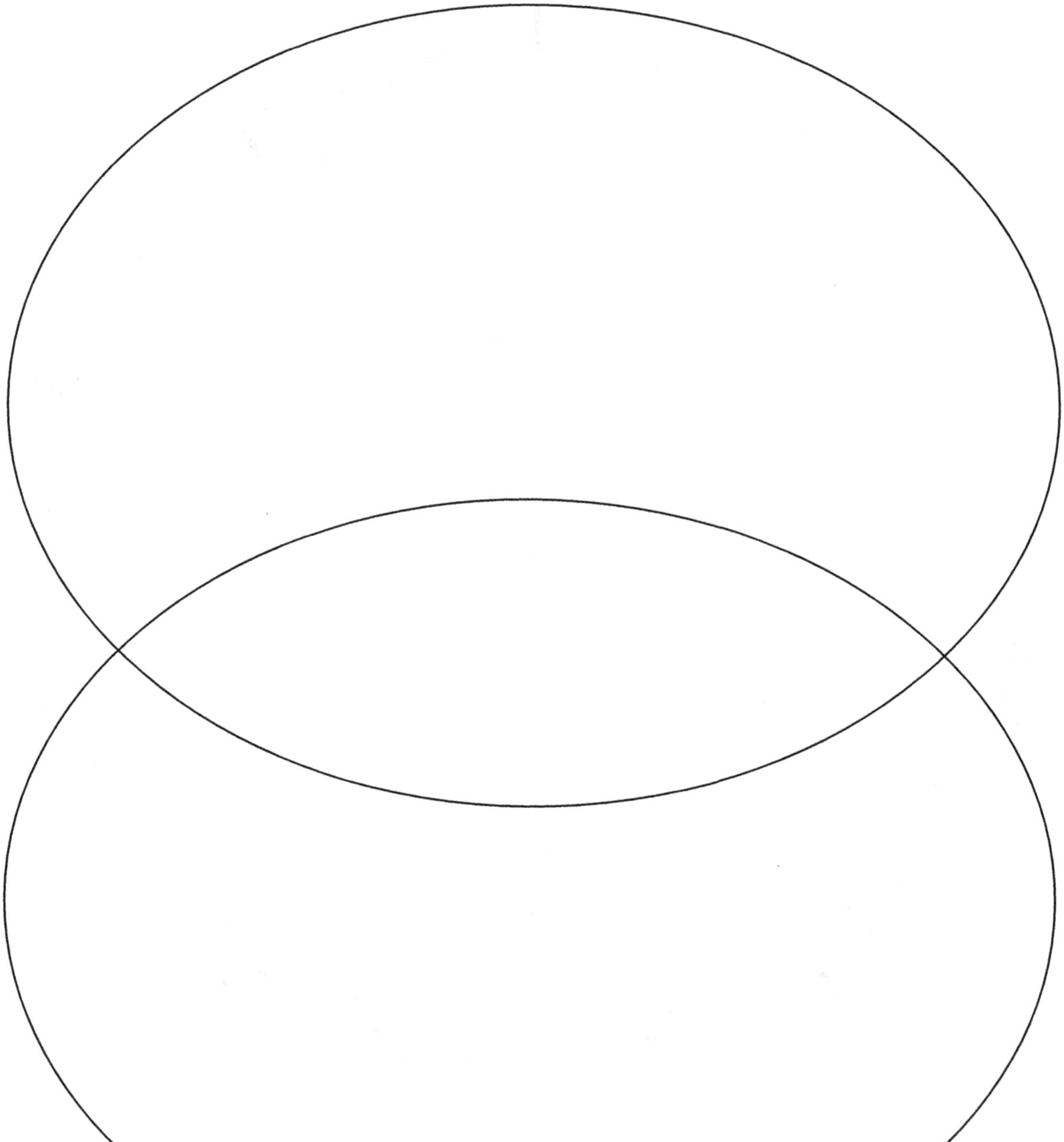

NAME DATE

53. Triple Venn Diagram Template

____

(title)

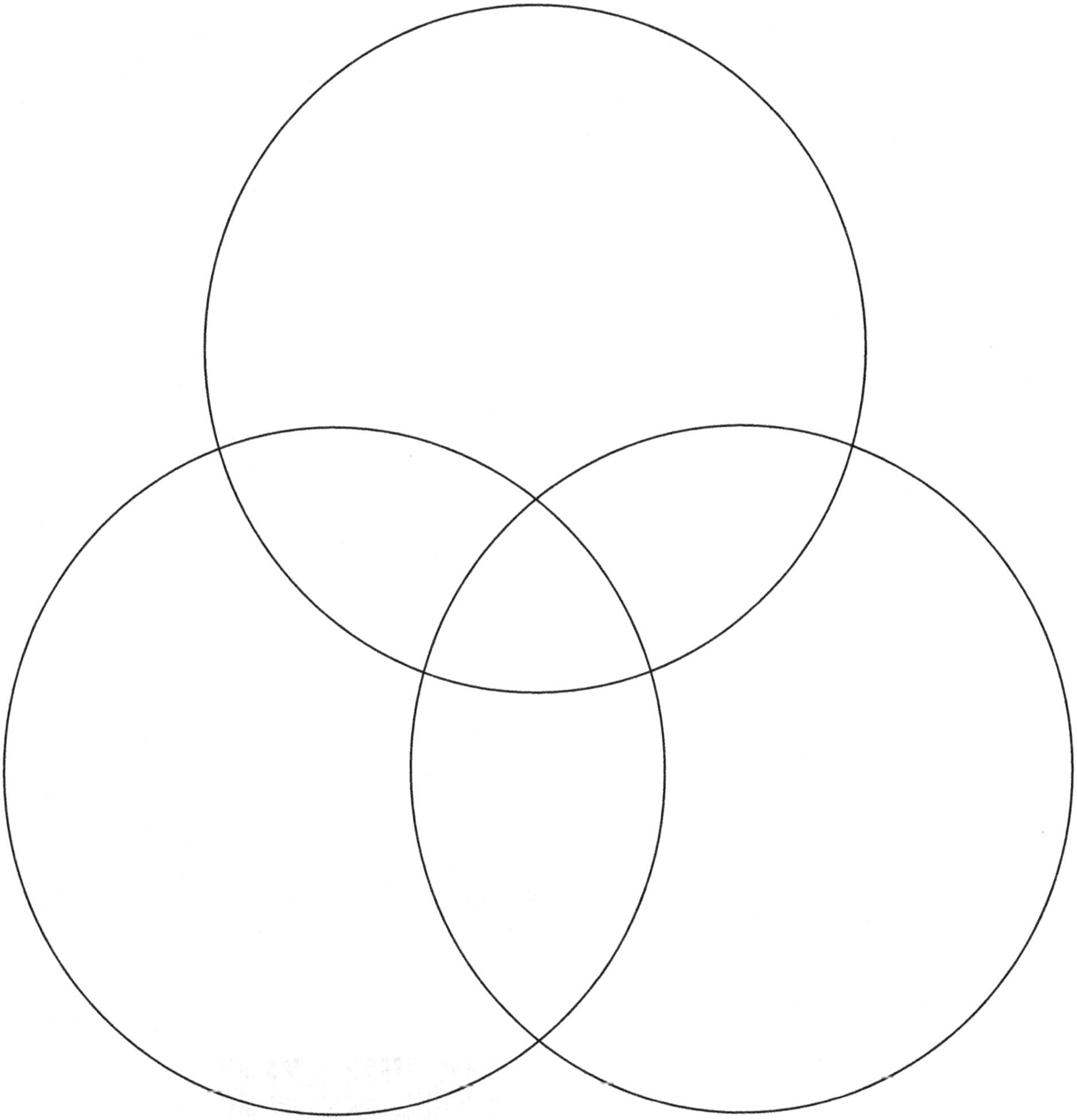

Made in the USA
Monee, IL
08 November 2026

67380781R00046